ELIJAH COME AGAIN

ELIJAH COME AGAIN

A PROPHET FOR OUR TIME
A SCIENTIFIC APPROACH TO REINCARNATION

ROBERT POWELL

LINDISFARNE BOOKS
2009

Lindisfarne Books

An imprint of SteinerBooks/Anthroposophic Press, Inc.

610 Main St., Great Barrington, MA 01230

www.steinerbooks.org

Cover and Book Design: William Jens Jensen
COVER IMAGE: Raphael, *The Transfiguration* (detail);
oil on wood, 405 × 278 cm; Pinacoteca, Vatican City

Elijah Come Again (2009) is published in commemoration of the hundredth anniversary of Rudolf Steiner's first presentation of the reincarnation example of the incarnations of the Old Testament prophet Elijah as John the Baptist, Raphael, and Novalis. This occurred in Munich, January 6, 1909, after Marie Steiner (then Marie von Sivers) made a remark to Rudolf Steiner (quoted here on page 70). The lecture was part of a Christmas celebration for members of the German Section of the Theosophical Society (known after 1913 as the Anthroposophical Society). In the lecture, Steiner spoke for the first time in a general members' lecture (independently of theosophical traditions) on the connections from incarnation to incarnation of identifiable historically personalities as reincarnations of a particular individuality. This majestic "karma revelation" was a world-historic event, taken up in this book one hundred years later. It is dedicated in honor of the eternal individuality of Elijah–John the Baptist–Raphael–Novalis, one of the great spiritual leaders of humanity.

LIBRARY OF CONGRESS CATALOGING-IN-PUBLICATION DATA

Powell, Robert, 1947–
  Elijah Come Again : a prophet for our time : a scientific
approach to reincarnation / Robert Powell.
    p. cm.
  Includes bibliographical references.
  ISBN 978-1-58420-070-3
  1. Reincarnation. 2. Reincarnation—Research. I. Title.
  BL515.P69 2009
  202'.37—dc22

                                          2008056136

Contents

Preface   ix

Introduction   xiii

1: Toward the Unveiling of the Mystery   1

2: Who is Elijah?   10

3: Concerning the Basics: John the Baptist–Raphael   18

4: Novalis   32

5: Raphael   45

6. Death Raphael–Birth Novalis   54

7. Chart Comparisons: Raphael–Novalis   61

8. St. Elisabeth–Raphael–Novalis   70

9. St. Elisabeth–Raphael–Novalis: Horoscope Comparisons   91

10. Phinehas and the Mystery of Individuality   103

11. Phinehas–Elijah–John the Baptist   120

12. The Mystery of Elijah and Lazarus–John   130

13. John the Baptist–St. Elisabeth–Raphael–Novalis   138

14. The Cosmic Journey of the Soul   151

15. The Two Witnesses   183

Afterword: Fulfillment of a Prophecy   194

Appendix 1: Truth and Error in Karma Research   197

Appendix 2: Astrological Reincarnation Chart Comparisons   202

Notes   217

Bibliography   232

Acknowledgements

With grateful thanks to the Sophia Foundation of North America and board members of the foundation, especially to cofounder Karen Rivers, for her work of steadfast support and encouragement, which has been a blessing for our work with Sophia and with the individuality who is the subject of this book, who can be regarded as an emissary for Sophia in our time.

My deep appreciation is extended also to the Institute for Astrosophy in Germany, to the board members, and in particular to Bernt Rossiwall and Arnold and Heidi Langen, for support received in pursuing scientific research into reincarnation and karma.

My heartfelt gratitude goes also to my wife, Lacquanna Paul, for her work on the manuscript of this book, helping to transform it into a highly readable and interesting karmic biography of the Elijah individuality. Through her endeavor this book is now in a form that is a worthy tribute to this great spiritual figure in the history of humanity.

Warm thanks also to Dianna Marsden, who typed the first draft of this manuscript, which has been gestating for many years since that time.

I would also like to thank publishers Chris Bamford and Gene Gollogly for taking on the manuscript and enabling this book to become available.

Finally, there are many others who have helped in one way or another to whom my gratitude and appreciation is extended, without naming everyone explicitly.

PREFACE

ULTIMATELY, EVERY AUTHOR MUST question the timeliness of a written work. So it is as I put the finishing touches to this manuscript, the repository of a timeless story representing more than thirty years of research into the question of reincarnation and human destiny.

This manuscript, for which the research was completed some time ago, resurfaces now with a sense of urgency for the telling of a destiny story that can serve as a reminder to the call for Love in the world. Through telling the story *Elijah Come Again,* we are reminded that humankind is involved in a mission to evolve evermore deeply into the capacity for love in the activity of the will.

Eastertide seems the appropriate time for tilling this fertile soil as the Elijah story resurfaces again, pressing out from the soil with the urgency of flowering bulbs in spring, as though undergoing a reawakening—bearing spiritual color and reaffirming radiance in a time of uncertainty in the world. This is a story whose seedlings were planted long ago, corresponding to Saturn's rhythm in the human soul.

Saturn time speaks always toward the development of human conscience, as a steadying force for morality in the right use of the will in serving the truth. Thus, there is a telling aspect to the timing of the initial draft of this manuscript in 2003, which was the year of the invasion of Iraq by US military forces—a time when the world soul, through the immediacy of the media reports of the war, was summarily engulfed in the emotional fallout and moral dilemma brought about by the US "shock and awe" bombing of Baghdad, which literally rained down fire from heaven, bringing devastation and chaos to the region of the Earth referred to as the "cradle of civilization," and awakening questions of moral conscience for most of the civilized world.

As an answering call for the truth, the story *Elijah Come Again,* emerges now in our time as a scientific approach to the unveiling of the mystery of human destiny, a story that is timeless in nature…yet, nevertheless, timely in the accounting of the unfolding destiny and mission of the Old Testament prophet Elijah—an individuality of such spiritual weight and stature as to call forth both rain and fire from heaven in the service of the evolution of humanity toward communion with the spiritual forces of Love and morality that live in the human heart as the abiding presence of the living Word.

My task now is to provide a living background for an exploration seen primarily through the lens of scientific research in order to bring the story of human destiny to the fore in the presence of a world that now has been seduced into believing that a sub-earthly force (nuclear power) will yield protection from the threat of war. This belief has fostered the accumulation of stockpiles of nuclear weapons, engendering great distrust and tension among nations, resulting in the building up of military strength and the real threat of nuclear war, the result of which—owing to the devastating consequences of radioactive fallout—would be an "air trial" for the whole Earth.

On the day of writing this preface, there surfaced a rumor of the threat of an air raid by US military forces seeking to destroy Iran's nuclear facilities. This is a curious reversal of the peaceful dismantling of nuclear arms that had been in process in the preceding years. Clearly something of an elemental fire nature has been ignited, and civilization as we know it is at risk. One might ask: Has humanity forgotten its true destiny? Is the world in need of an awakening call from the spiritual world to set things right? *"Elijah does come first to restore all things"* (Mark 9:12). These are words of Christ, and I believe this story to be an ongoing unfolding of Elijah's mission toward the peoples of the Earth. From the fall of Babylon to the invasion of Baghdad, there is the continuity of a "work in progress"—an Old Testament story continuing on into our time is seeking resolution.

As I look out toward the horses grazing peacefully in the pasture beyond our village garden, I ponder the mystery of an earlier time in my own biography, when my wife and I sang in a choir—comprising one hundred singers (from Weilheim, Bavaria, and other neighboring towns)—Mendelssohn's great oratory *Elijah,* accompanied by a full orchestra. This was on October 26, 1991. Now, in looking back, I am reminded that 1991 was the year of the first Gulf War.

Mendelssohn was born in Hamburg, Germany, on February 3, 1809, to parents of Jewish descent. He converted to Christianity and received the inspiration for writing *Elijah* in 1837 at the age of twenty-eight. Requiring nine years of gestation, the oratory was first performed in Birmingham, England, on August 26, 1846. Mendelssohn died the following year in Leipzig on November 4, 1847. In his magnificent oratory *Elijah,* there is something living for the future. Might this be a testimony to Elijah's continuing presence as a guiding force for humanity? To answer this question we must look at the developing consciousness that was awakening during this time period.

Rudolf Steiner indicated that 1841 marked the beginning of the *war in the heaven*, described in chapter 12 of the Book of Revelation, with the Archangel Michael casting down the "spirits of darkness" in preparation for his becoming the *Spirit of the Age* in November 1879—for the age that lasts until approximately 2234.[1] Thus Mendelssohn's *Elijah*, composed during the time period when the *war in heaven* (1841–1879) commenced, came as a sounding of prophetic grace for the times ahead.

It is striking to recall that its first performance came to my native England, and correspondingly to the English-speaking world, two years before the publication of Karl Marx's *Communist Manifesto* in London in 1848, which brought a revolutionary spirit to the air at a time that was already full of unrest—a time characterized by socialist uprisings in various European cities, such as the Dresden uprising of 1849, in which the composer Richard Wagner participated, together with his Russian friend, the anarchist Mikhail Bakunin.

The sound of a hundred voices echoing in the concert hall in the Bavarian town of Weilheim, singing out the words of the *Elijah* oratory from the Old Testament, live on in me not only as a testament to the past but also as an awakening call to Elijah's future destiny and his mission as a heraldic presence, a living accessible reality for the bringing to birth of moral conscience—a shepherding presence or voice in the wilderness calling *metanoia* ("change your ways"), addressing human consciousness in today's distraught world.

INTRODUCTION

REGARDING THE SCIENTIFIC INVESTIGATION of reincarnation—
also known as "karma research"—it is appropriate to introduce
this book with a few words of a personal nature, to describe my relationship to the activity of karma research.

Nowadays there are many books on reincarnation, several of
which describe people's real or imagined experiences from previous incarnations. Among this voluminous literature there is one
author who, through my own scientific research, I have found to
be completely reliable—Rudolf Steiner.[2] This book builds upon his
indications concerning reincarnation and karma, and at the same
time extends them in a reliable, scientific way, as will be shown.
Nevertheless, the *reader* may justifiably ask: How might findings of
reincarnation research be considered reliable? This is the question
that arose within me when reading Steiner's reincarnation findings.
Moreover, it is this *living question* that has led me to the research
findings indicated in this book.

In the course of discovering the validity of Steiner's reincarnation indications, it became clear to me that it is precisely through the
science of reincarnation and karma that a *new* science of the stars
can come into being, and that this could be undertaken as a spiritual task. Through Rudolf Steiner, this spiritual task was awakened
within me in 1972.

As a scientist and spiritual seeker, it is essential to be concerned, like Rudolf Steiner, with truth and with truth alone. In
the course of years of research, I have found it possible to test the
validity of reincarnation statements in a reliable way. Therefore,
it can be stated that, just as it is possible to know beyond any
shadow of doubt that the Pythagorean theorem is true and prov-
able (that the sum of the squares of the lengths of two sides of a

right triangle is equal to the square of the length of the hypotenuse), so it is also possible to know if a reincarnation statement is true or not. For example, Rudolf Steiner made the statement that the Old Testament prophet Elijah reincarnated as John the Baptist. Confirmation of this reincarnation statement can be found in the *Gospel of St. Matthew*:

> And the disciples asked him, "Then why do the scribes say that first Elijah must come?" Jesus replied, "Elijah does come, and he is to restore all things; but I tell you that Elijah has already come, and they did not know him, but did to him whatever they pleased. So also the Son of man will suffer at their hands." Then the disciples understood that he was speaking to them of John the Baptist. (Matthew 17:10–13)

Thus, it is possible to know with certainty that the statement "Elijah reincarnated as John the Baptist" is true. But how is this possible? In this example, we have the word of Jesus Christ that it is true—he who said of himself: "I am the way, the truth, and the life." If we accept this statement that he, the Logos, is THE TRUTH, then ultimately it is through Christ that we know the truth, or through someone or some being in his service, or through his feminine counterpart, Sophia (whom we will consider later in this book). As far as it is possible to judge upon the basis of his life's work, Rudolf Steiner was in the service of both Christ and Sophia (his teaching is called "Anthropo-*Sophia*," or *Anthroposophy*). I believe that what we know of spiritual truth through Rudolf Steiner is verified, in the last analysis, by Christ and Sophia.

Regarding the connection of Sophia (whom the Egyptians called Isis[3]) to karma research, let us contemplate the following words from Rudolf Steiner's "karma lectures":

> A German romanticist once had the courage to think differently from all others about the famous saying of Isis: "*I am that which was, that which is, and that which is to come; and my veil no mortal has yet lifted.*" To which the German romanticist replied: Then we must become immortal; that we may lift the veil![4]

That German romanticist is none other than Novalis. He knew, of course, Schiller's ballad, "The Veiled Image at Sais," published in 1795. This ballad is about a youth who is on a journey in Egypt and who comes to the city of Sais, where the great image of veiled Isis was located, with an inscription that translates as that saying attributed to Isis and quoted by Rudolf Steiner. In Schiller's ballad, Isis herself warns the youth, "No mortal lifts this veil until I myself do so." Despite knowing of the punishment awaiting him, the youth makes an attempt, slipping into the Isis temple one night and lifting the veil. He pays dearly, losing his life. In contrast, Novalis wrote in his *Novices at Sais:* "And if no mortal, according to the inscription there [at Sais], lifts the veil, then we have to seek to become immortal."[5] Elsewhere Novalis wrote,

> One succeeded—he lifted the veil of the goddess at Sais.—
> But what did he see? He saw—miracle of miracles—himself.[6]

This is a fitting reflection for the following statement by Rudolf Steiner, which brings the lifting of Isis' veil into connection with the activity of research into the mysteries of reincarnation and karma:

> Perception of the karma of an individual human being, or even of one's own karma, requires the right attitude, the right mood of soul. The whole study of karma is profaned if this study is pursued in the attitude of mind arising from our modern education and civilization. The mood in which all teachings about karma should be received is one of piety, of reverence. Whenever one approaches a truth relating to karma, one's soul should feel as though part of the veil of Isis is being lifted. For in truth it is karma that reveals, in a way most intimately connected with human life, what Isis was—the being designated outwardly as: "I am that which was, is, and will be." This must be the attitude of soul in all study of human karma.[7]

Here a connection of Isis-Sophia to the realm of karma research is expressed. In our study of karma, with every truth that is revealed, we should have the feeling that "part of the veil of Isis is being lifted." Isis-Sophia is the being referred to in the Book of Revelation as the "Bride of the Lamb," whereas the Lamb is, of course, Christ. In the tradition of Christian esotericism, Christ oversees the unfolding of human destiny. Thus Christ is called the "Lord of destiny," or the "Lord of karma." Similarly, Sophia, as the Bride of the Lamb, also oversees the unfolding of destiny together with Christ. Taking this into account makes it is possible to understand the connection of Isis-Sophia with the realm of karma research.

If we accept their role as the overseers of destiny, then ultimately it is Christ and Sophia who enable us to know the truth of a reincarnation statement with certainty. Owing to Christ's renewed activity beginning in the twentieth century (his coming in the etheric realm, as described in detail elsewhere [8]), a new revelation of Christ is now taking place. This revelation essentially has two sides; one is a new revelation of nature, whereby—through Christ—perception of the elemental beings and etheric forces at work in nature is becoming increasingly possible. The other side is a new revelation of the mysteries of reincarnation and karma. This is the *informing* circumstance behind increasing numbers of people being led to an experience of their former incarnations on Earth and of their karmic relationships with other people.

Simultaneously, there are serious dangers associated with this, as alluded to in Schiller's ballad "The Veiled Image at Sais," with the forewarning of Isis: "No mortal lifts this veil, until I myself do so." Cases of delusion and becoming less than fully grounded in the reality of one's present incarnation are all too common, causing one to miss opportunities for accomplishing one's true karmic task.

As a guiding light toward truth, let us consider Valentin Tomberg's words relating to the new power of clairvoyance called forth by the revelation of Christ in the etheric:

> The new natural clairvoyance will reveal itself among human beings in two ways. Some people will see the spiritual at work in nature—will recognize how the elements are an expression of those spiritual mysteries that have flowed into the Earth.... The mystery of the evolution in nature will become ever more apparent to these individuals. In another group, a different capacity will emerge—a karmic seeing, the beholding of the karmic past. They will behold the karma called forth by various deeds. Rudolf

Steiner also spoke about this. I should add that the mode of seer-ship of such individuals will consist in their being endowed with a clairvoyance that investigates the karma of humanity. They will remember their own karmic connections, then those of other people, and so on. [9]

This exactly describes how it came about that I was inspired to investigate the karma of humanity. My initial awakening took place through a spiritual encounter with Rudolf Steiner, whom I recognized as an "inner plane master"—a fully conscious being, able to move about in space and time, and with whom one could converse spiritu-ally. I began to study his books, beginning with the first four volumes of his lecture series entitled *Karmic Relationships*. During that period (1972), I also began to read the writings of Willi Sucher (1902–1985), the pioneer of *astrosophy*. I recognized in Sucher's scientific approach to astrosophy a potential scientific tool for the investigation of karma. The *development* of astrosophy *in this direction* was the task that faced me, and this led me many years later to write the two volumes entitled *Hermetic Astrology*. [10]

Before elaborating upon this, however, let us return to the quote by Valentin Tomberg. He states: "They will remember their own kar-mic connections, then those of other people, and so on." This is pre-cisely what took place in my life in 1977, at the age of thirty. First I was guided to discover my own previous incarnations, and then those of friends and acquaintances. This continued, not in a spectacular way, but in a slow and steady way, through subsequent meetings with people. These were the initial inspirations by which I set about apply-ing the new star wisdom (pioneered by Willi Sucher) to the findings of karma research. What I quickly came to realize was that the method of applying astrosophy to the findings of karma research contained the potential for discovering the *foundations* of a completely new star wisdom. I would like to make the significance of this clear by way of an analogy drawn from the realm of astronomy.

The Danish astronomer Tycho Brahe (1546–1601) dedicated more than twenty years of his life to cataloging the fixed stars and making exact observations of the movements of the planets against the back-ground of the fixed stars of the zodiacal constellations. Having collected all these observations, he realized that he needed someone who could analyze them and discover the laws of planetary motion. It was destiny that led the German mathematician/astronomer Johannes Kepler (1571–1630) to work with Tycho Brahe. On the basis of Tycho's observations, Kepler discovered the three laws of planetary motion that led to a completely new astronomy, entirely reforming the old astronomy.

By way of analogy, like Tycho Brahe, I have been collecting observations for over thirty years and have amassed a considerable collection of observations/findings of research into reincarnation and karma in relation to the development of a new star wisdom (astrosophy). This book—going beyond the two volumes titled *Hermetic Astrology*—comprises an initial sharing of my findings. I say "initial" because there is much more to be done. My work of thirty years has only begun to plumb the depths of this vast new area of research. This is one of the primary reasons for writing this book, in the hope that there will be others who will achieve with respect to the findings/observations of karma research, what Johannes Kepler accomplished with regard to the observations of Tycho Brahe.

Just as four hundred years ago the *astronomia nova* (new astronomy) was born through Kepler's analysis of the observations of Brahe, so now I believe an *astrologia nova* (new astrology) is coming to birth through an analysis of the discoveries uncovered through investigating authentic reincarnation examples in the light of star wisdom (astrosophy). This book—like Rudolf Steiner's series of volumes entitled *Karmic Relationships*—contains authentic reincarnation examples. The horoscopes of these reincarnation examples, where possible, are also presented.[11] This book will indicate the scientific and spiritual means for testing their validity.

A start has been made with the analysis of the examples presented in this book, yet this is a new area of scientific research that needs to be developed further. I believe that once the findings of astrological reincarnation research have been assimilated, a complete reformation of traditional astrology will inevitably take place. And when the new astrology is established, there will be a similar feeling in looking back upon traditional Western astrology that modern astronomers have in looking back upon the old geocentric astronomy.

This book is a work that stands at the threshold between science and spirituality. It offers an indication to humanity for a way into the future. It accepts the scientific method, yet applies it to a completely new domain—that of reincarnation and karma. The results are revolutionary, in that, through the analysis, methods for use in future astrological research relating directly to human destiny have been discovered, thereby reinstating the original purpose of astrology.

Many volumes could be written about the spiritual significance of Elijah, of whom the Lord said to the prophet Malachi, the last of the Old Testament prophets: "Behold, I will send you Elijah the prophet, before the great and terrible day of the Lord comes" (Malachi 4:5). The purpose of this book is to contemplate the Elijah series of incarnations with the goal of laying the foundation for a new "science of the stars" as the "science of karma."[12] There is a golden thread of connection running through this work. Although a scientific mode of presentation demands a certain degree of rigor, this book is nonetheless written not only as a textbook for the new science of the stars as a science of karma, but also as a source of inspiration for the reader concerning the individuality of Elijah, who "is to lead humankind past the great crisis in which it is involved."[13]

This image depicts the upper part of Raphael's painting of the Transfiguration. This was the last painting that he was working on at the end of his life. Because of his earlier incarnation as the prophet Elijah, it is a scene with which Raphael was connected existentially. The image shows, from left to right, Moses, Christ, and Elijah. Moses and Elijah are the two witnesses mentioned in Revelation 11, as discussed in this book, which is dedicated to one of these two witnesses.

Toward the Unveiling of the Mystery

Two "Laws" of Reincarnation and Karma Research[14]

A S PART OF THE applied method of research presented in this book,
there are two fundamental ideas to which we will need to refer.
These two fundamental ideas were born of the results discovered
through the astrological reincarnation research presented in *Hermetic
Astrology* volumes I and II, and have continued to serve as a method
of analysis. In addressing the basic principle of astrological reincarnation research, we begin with the comparison of the horoscope of
birth with the horoscope of death from the previous incarnation. This
principle was formulated by Rudolf Steiner:

> When a person passes through the gate of death he dies under
> a certain configuration of stars. This configuration is signifi-
> cant for his further life of soul because it remains there as an
> imprint. In his soul there remains the endeavor to enter into
> the same configuration at a new birth, to do justice once again
> to the forces received at the moment of death. It is an interest-
> ing point that if one works out the configuration at death and
> compares it with the configuration of the later birth, one finds
> that it coincides to a high degree with the configuration at the
> former death.[15]

The principle set forward by Rudolf Steiner in the above statement
served as the starting point for the research presented in volume I
of *Hermetic Astrology* wherein the reincarnation examples given
by Rudolf Steiner, where dates of birth and death are known, were

analyzed by comparing the birth horoscope with the death horoscope from the previous incarnation.

This endeavor led to the discovery or to the uncovering of what I
refer to as the first two rules or "laws" of astrological reincarnation.
These are not to be considered as "hard-and-fast laws" but rather as
flexible rules, since they do not apply in every case. Perhaps this phenomenon of flexible rules of astrological reincarnation could be considered in relation to the Uncertainty Principle formulated by Werner
Heisenberg in 1926.

The First "Law" of Astrological Reincarnation

Keeping the above commentary in mind, the first rule of astrological reincarnation states that the angular relationship between
the Sun and Saturn at death recurs—in the same way or in a metamorphosed way—at birth in the next incarnation.[16] According to
Rudolf Steiner:

> This relationship of Saturn to the Sun expresses itself in the
> human being such that first the "I" [Sun] comes into a corre-
> sponding relationship to the astral body [Saturn] and, however,
> that the astral body is membered in the right way into the entire
> human organization.[17]

For example, the angular relationship between Sun and Saturn
at Rudolf Steiner's birth was 179° and at death in his previous incarnation it was 176°; in both cases this angle is called an "opposition"—with the positions of Sun and Saturn appearing approximately opposite one another in the 360° circle of the zodiac, where
an angle of 180° would signify exact opposition. This can be readily
seen in the following horoscope example, whereby the birth horoscope of Rudolf Steiner is cast in the center surrounded by the horoscope of death from his prior incarnation cast in the outer circle
(see next page).[18] As will emerge in the following pages, for the
purpose of astrological reincarnation research, the horoscopes are
sidereal, based on the original zodiac of the Babylonians, Egyptians
and Ancient Greeks, rather than the tropical horoscopes of modern
Western astrology.[19]

Comparison Chart

Outer - Geocentric	Inner - Geocentric
Death of (St.) Thomas Aquinas	Birth of Rudolf Steiner
At Fossanova, Latitude 40N50', Longitude 14E15'	At Kraljavec/Yugoslavia, Latitude 46N22', Longitude 16E39'
Date: Wednesday, 7/MAR/1274, Julian	Date: Monday, 25/FEB/1861, Gregorian
Time: 3:0, Local Time	Time: 23:25, Local Time
Sidereal Time 14:27:25, Vernal Point 15 ♓ 22'33"	Sidereal Time 9:47:41, Vernal Point 7 ♓ 11'52"

House System: Placidus, Zodiac: Sidereal SVP
Aspect set: Conjunction/Square/Opposition

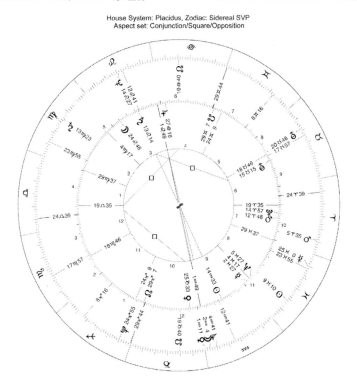

Example of the first "law" of reincarnation:

geocentric—Death Thomas Aquinas (outer) / geocentric—
Birth Rudolf Steiner (inner)

The outer circle of this comparison horoscope gives the planetary configuration at the death of Thomas Aquinas. Here we see Saturn (13♍23) in Virgo is opposite the Sun (9♓10) in Pisces, therefore the angle between Saturn and Sun is 176 degrees, rounded to the nearest degree. Similarly, in the inner circle, which gives the planetary configuration at Rudolf Steiner's birth, the Sun (14♒33) in Aquarius is opposite Saturn (13♌14) in Leo, signifying that the angular relationship between Sun and Saturn is 179 degrees, rounded to the nearest degree.[20]

Interchange

Whether or not the first rule regarding the angular relationship between Sun and Saturn holds between death in one incarnation and birth in the next incarnation, there can sometimes occur an *interchange* of the positions of Sun and Saturn either in conjunction or in opposition to the previous horoscope alignment, i.e. where the Sun's location at birth is at the same (conjunction) or opposite (opposition) zodiacal location to Saturn in the previous death horoscope or vice versa. This interchange of the locations of Sun and Saturn from one incarnation to the next, whether in conjunction or in opposition, is called an "interchange." In volume I of *Hermetic Astrology* I refer to the fulfillment of this interchange of zodiacal location of Sun and Saturn as the extended first "law" of reincarnation.

The Second "Law" of Astrological Reincarnation

Through the ongoing work of astrological reincarnation research, after the discovery of the first "law" of astrological reincarnation in 1976, a second rule of astrological reincarnation was discovered in 1982. This second "law" I believe bears the potential to completely revolutionize astrology. The implications of this discovery, when fully grasped, are so far-reaching that they could ultimately help to restore astrology to its original role as an indicator of an enlightened path for the reincarnating soul, providing a picture of how the stars serve in shaping and shepherding the soul's journey through aeons of time. This second "law," which holds only in terms of the sidereal zodiac, can be stated as follows:

> The sidereal zodiacal position(s) of heliocentric Mercury and/ or heliocentric Venus at birth in one incarnation tend(s) to align with the sidereal zodiacal position(s) of heliocentric Mercury and/or heliocentric Venus at death in the preceding incarnation.[21]

As an example of this second "law," let us again consider the sidereal horoscope of Rudolf Steiner. At Rudolf Steiner's birth, heliocentric Mercury was at 25° Taurus, aligned with the heliocentric position of Mercury (27° Taurus) at death in his previous incarnation.[22]

*Heliocentric alignment: Mercury from Death to Birth: 27°
Taurus (Death) — 25° Taurus (Birth)*—see next page.

The Sidereal Zodiac

Recalling that *heliocentric* refers to the perspective from the Sun that is the domain of the higher self, this second rule offers scientific proof to the significance of two important elements of which the traditional practice of Western astrology—with few exceptions—is unaware:

> First, for the purposes of karma research, confirmed through thirty years of application using Rudolf Steiner's indications, the sidereal zodiac has proven to be *authentic* as an accurate astrological template. This did not prove to be true with the tropical zodiac of traditional Western astrology, since the alignments (such as the one referred to here between h-Mercury at death and h-Mercury at subsequent rebirth) between planets from one incarnation to the next show up in the sidereal zodiac and not in the tropical zodiac.

Sidereal means "of the stars." This term refers to the *original* zodiac used by the Babylonian, Egyptian, Greek, and Roman astrologers. These were stargazing cultures that observed the movements in the starry heavens and thus were the early gatherers of the wealth of cosmic knowledge passed down in the science of astrology. It is the sidereal zodiac—or a close variant thereof—that is used to the present day in Hindu/Vedic astrology. In contrast to Vedic astrology, traditional Western astrology utilizes the tropical astrology practiced by Arab astrologers. This is based on the tropical zodiac that was introduced into the West in the middle of the twelfth century c.e., when Arabic astrological texts were translated into Latin. [23]

Let us consider now what comprises the difference between the two systems of thought regarding the zodiac as an accurate template for the casting of horoscopes. The sidereal zodiac divides the circle of zodiacal constellations into twelve equal stellar signs, i.e. twelve equal-length constellations, each thirty degrees long (see page 8). This provides a system in which the actual location of stars provides the

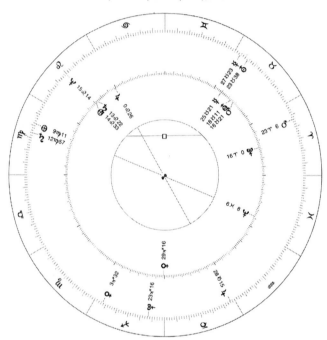

Comparison Chart

Outer - Heliocentric	Inner - Heliocentric
Death of (St.) Thomas Aquinas	Birth of Rudolf Steiner
At Fossanova, Latitude 40N50', Longitude 14E15'	At Kraljavec/Yugoslavia, Latitude 46N22', Longitude 16E39'
Date: Wednesday, 7/MAR/1274, Julian	Date: Monday, 25/FEB/1861, Gregorian
Time: 3: 0, Local Time	Time: 23:25, Local Time
Sidereal Time 14:27:25, Vernal Point 15 ♓ 22'33"	Sidereal Time 9:47:41, Vernal Point 7 ♓ 11'52"

Zodiac: Sidereal SVP
Aspect set: Conjunction/Square/Opposition

Example of the second "law" of reincarnation:

Alignment of ♄-Mercury – Birth Rudolf Steiner (inner)
with ♄-Mercury – Death Thomas Aquinas (outer)

observable background to the movement of the planets. In contrast, the tropical zodiac is defined to commence with the vernal point, where the Sun is located on March 20/21, and is thus of a calendrical nature, pertaining to the cycle of the seasons in the northern hemisphere. Due to the backward movement of the vernal point through the zodiac, regressing 1 degree every seventy-two years, the tropical (calendrical) description of the zodiac no longer relates to the actual astronomical location of the constellations of stars referred to by astronomers as the zodiac. Thus one can see how—for the purposes of astrological research—the tropical (calendrical) zodiac gives a picture unrelated to the actual zodiacal background of the location of the planets. Due to the discovery that the planetary alignments from one incarnation to the next are only apparent against the background of the sidereal zodiac, which is related to the astronomical location of the stars, the horoscopes presented in this book—unless otherwise stated—are in terms of the sidereal zodiac.

Now we come to another far-reaching implication of the second "law" of astrological reincarnation research:

1. The second "law" points to the discovery that the heliocentric horoscope (the horoscope from the vantage point of the Sun) is highly significant in addition to the geocentric horoscope of traditional astrology. This opens up a new dimension, expanding beyond the Hindu/Vedic astrology, that represents a continuation of the ancient Babylonian stargazing astrology, in which *only* the geocentric sidereal horoscope (the horoscope from the perspective of the Earth) is considered.

2. The extended first "law" and the second "law" of astrological reincarnation indicate that the geocentric horoscope *and* the heliocentric horoscope of birth need to be taken into consideration, each cast in terms of the sidereal zodiac. [24]

These two findings, revealed through astrological investigation into the inherent truth of Rudolf Steiner's reincarnation indications, are actually both implicit in the second "law." Taken seriously, they necessitate a completely new approach to astrology. At the present point in time, at the start of the third millennium of the Christian era, the two "laws" of astrological reincarnation research that have been *uncovered* through the results of empirical research are for the most part unknown in the general practice of astrology in the West. Similarly, the discoveries made possible following from their uncovering remain *veiled* to present-day consciousness.

If the purpose of astrology is to serve as a guide and indicator of destiny and soul development, then a *reliable* science based upon the *actual* location of the stars has proven through over thirty years of research to be the observation of and the adherence to the astronomical locations of the planets against the background of the sidereal zodiac.

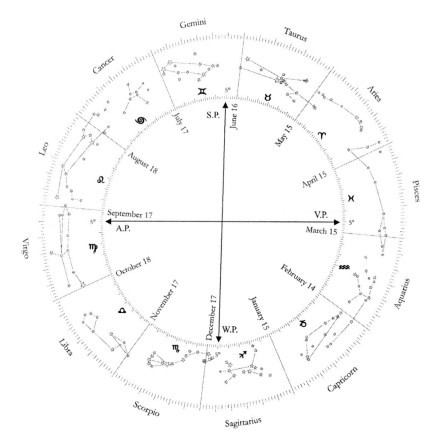

Figure 1: The sidereal zodiac—
Dates of the Sun's ingresses into the signs of the zodiac

This fundamental truth is obvious if one undertakes a new approach toward the practice of astrology based on the "laws" of astrological reincarnation that have been discovered so far. One of the primary purposes for writing this book is to help facilitate the discovery of further "laws" by providing a "revealing body" of empirical research and thus to further a renewal in the practice of astrology—leading not only to an *Astrologia nova* ("new astrology") but ultimately to an *Astrosophia,* the study of the wisdom of the stars as a revelation of Sophia's body of wisdom.

This brings us to the subject of the astrological survey in this book, calling attention to the great Elijah individuality of whom Jesus spoke the words, "Elijah does come, and he is to restore all things" (Matthew 17:11).

In the service of Sophia, we now embark upon a scientific inquiry and study into the destiny stream of the great Elijah individuality, referred to in *The Most Holy Trinosophia* as the "third spiritual teacher" of our time.[25] This third teacher *lives* as an ever present source of inspiration, guiding souls to the great School of Divine Wisdom or *School of Sophia*. The connection that can be found with the third teacher is a purely inner one, as she is not at all active in the world in an outer way. She acts—on a spiritual-soul level—as a center for the mediation of spiritual impulses, as did the Virgin Mary for the apostles and disciples. Two thousand years ago the Virgin Mary was at the center of the circle of apostles and disciples and acted as a vessel for Sophia, through whom the descent of the Holy Spirit was accomplished at Pentecost.[26] The third teacher has an analogous role now, acting as a vessel for Sophia in our time. The purpose of this book is to pay homage to this great teacher of humanity by focusing upon her past incarnations as a fiery prophet, the herald of Christ, a saintly nun, a great artist, and an inspired poet.

Through the astrological reincarnation research presented in this work, the reader is introduced to the cosmic background and wisdom inherent to the sequence of the incarnations of this individuality. In this way the reader is brought into a *living* imagination of the karmic biography of the third teacher. To those readers who are interested more in the homage paid to this great teacher, my advice is to pass over the sections that concentrate purely on astrological research comparisons and concentrate upon the unfolding of the biographical story that is interwoven throughout the text. The interweaving of the research and the biography reveal a mystery belonging to our time— with the purpose to serve as an inspiration toward the beginning of a true relationship with the heraldic wisdom of the coming of Christ in the etheric body of the Earth. Upon reading this book, perhaps you will agree that the results of the research presented here into the question of the truth of reincarnation have *unveiled* mysteries central to the evolution of humanity. *"Elijah does come, and he is to restore all things"* (*Matthew* 17:11).

2

WHO IS ELIJAH?

WHO IS ELIJAH AND what significance does he carry for the moral dilemma arising in our modern world? Most of us from the Judeo-Christian tradition and related faiths recognize Elijah as the Old Testament prophet who was taken up into heaven in a chariot of fire. We may recall also that Elijah is mentioned in the New Testament by Jesus as his sole indication of reincarnation, alluding to John the Baptist as the reincarnated Elijah, a reference that has been discussed earlier in the introduction.

In this chapter we shall call to mind some of the highlights from the life of Elijah, as they are remarkably significant to our times and reveal a golden thread of connection between the Old and New Testaments. In this way we are able to discover a background and testament for our own time in history. The events in the life of Elijah serve as a revelation indicating that what we are facing now is a continuing development of an earlier time, involving the karmic echoes from past actions coming either from a fallen nature or stemming from the karmic results of what is carried forward of a higher nature. Whereas the former behaviors find their source of influence from counter-forces of a destructive nature, the latter reflect our true nobility as human beings—worthy of being regarded as the crown of creation here on Earth.

The images that are held sacred in our modern world serve as icons for what we have come to believe is important and life-bringing, and many of these images are clearly recognizable in the life of Elijah.

"Partial histories are thoroughly impossible—each history must be a world history, and only in relation to the whole history is the historical treatment of a particular situation possible." [27] These words of insight coming from Novalis as a pearl of wisdom recorded for posterity are enlightening as a guide toward our approach to the *resounding* nature of historical considerations. Let these words be our guide as we begin our inquiry into the life of the Old Testament prophet Elijah for a better understanding of our world today.

Were he alive now, Elijah would be seen in our modern age as a zealot, meaning that he was fearless in his capacity to speak for the truth. The word *capacity* is used here because to be a prophet signifies a capacity to commune with the spiritual world and to be a spokesman for a higher reality. This signifies that Elijah's words bore a primal quality and strength so as to be able to call upon and enliven elemental powers to deliver forces of such primal strength as to awaken responses within the human soul. Following upon this, we could say that Elijah was a prophet of the Word and that his speaking bore an elemental force. From the events in his life, we can already recognize that Elijah served even in his day as a prophet for the coming of the healing redemption of the Christ and that his life bore a signature of cosmic dimension.

In the Orthodox Church, Elijah is revered as a powerful intercessory for healing. I like to think of him in this way, as an intercessory of healing. From these words of Novalis:

> In most religious systems we are regarded as members of the Godhead. Hence if one doesn't obey the impulses of the whole...if one doesn't even intentionally act against the law of the whole, but rather only follows one's own course, and will not be a member, then from the Godhead emanates a healing power, and either the wound is healed, or it will be utterly excised.[28]

When considered in relation to the being and life of Elijah, we can surmise that Elijah had an understanding of this primal corrective force and was in his day its spokesman.

With these thoughts as an introduction, let us begin to consider the events from the life of Elijah as recorded in the Old Testament. He was called Elijah the Tishbite or Thesbite, and was of the inhabitants of Galaad. Tishbite, meaning "that makes captive," most probably refers to his place of birth, assumed to be Teshbe. Galaad refers to the land of Gilead, the region where he was born, a mountainous region to the East of the River Jordan, lying partly in present-day Jordan.[29]

Elijah seemed to have had little concern for his physical well-being when in the face of what he deemed unrighteous. He chose to speak out the truth in a most forceful way. According to the Old Testament, the people of Israel had been chosen to bear a particular mission—to serve the one God, forsaking the idolatry of the pagan gods—and to live their lives according to the laws of Moses, the sacred law of righteousness set down in the Ten Commandments. With this in mind, we can imagine Elijah approaching King Ahab, the king of Israel.

Ahab was not living according to the law. He had taken a pagan wife, Jezebel, and through her he had submitted to idolatry, bringing a reversal to the progress of the developing consciousness of the people of Israel:

> And Ahab...did evil in the sight of the Lord above all that were before him.... He took also to wife, Jezebel, daughter of Ethbaal, king of the Sidonians. And he went and served Baal, in the temple of Baal, which he had built in Samaria; and he planted a grove; and Ahab did more to provoke the Lord, the God of Israel, than all the kings of Israel that were before him.
> (I Kings 16:30–33)

Baal worship was not a good thing for the people of Israel. It posed a definite threat to their survival, something that is discussed in greater detail in the chapter on Phinehas, who was an earlier incarnation of Elijah. Ahab's behavior inflamed the inner fire of the holy prophet, and Elijah rebuked Ahab. The Old Testament describes Elijah in words similar to those applied to him in his later incarnation as John the Baptist. Dressed in nothing but animal skins laced together with bits of leather, he approached King Ahab with a torrent of rebuttal, rebuking him and warning him that a drought would occur because of Ahab's wickedness.

Ahab loved his grove and wanted more. He was weak and succumbed to the wiles and suggestions of his wife Jezebel. She had seen to it that the owner of a local vineyard, Elijah Naboth, was murdered in order that her husband might seize the land, which he much coveted and could not otherwise obtain. Ahab made war on other tribes and kingdoms as well and was responsible for much death and devastation. Elijah abhorred all that Ahab represented.

Needless to say, Elijah's rebuke was not pleasing to Jezebel, and the drought that ensued was thought to be brought on by a curse from the prophet himself, so that it was now Elijah the Tishbite's life that Jezebel sought. Thus it was necessary for Elijah to flee into the desert:

> And Elijah the Tishbite, of the inhabitants of Galaad, said to Ahab: As the Lord liveth, the God of Israel, in whose sight I stand, there shall not be dew nor rain these years, but according to the words of my mouth. And the word of the Lord came to him, saying: Get thee hence, and go towards the east, and hide thyself by the torrent of Carith, which is over against the Jordan. And there thou shalt drink of the torrent: and I have commanded the ravens to feed thee there.... And the ravens brought him bread and flesh in the morning, and bread and flesh in the evening; and he drank of the torrent. (I Kings 17:1–4, 6)

Because of the drought, the brook eventually dried up and the Lord guided Elijah further:

> Arise, and go to Sarephta of the Sidonians, and dwell there: for I have commanded a widow woman there to feed thee... He arose, and went to Sarephta. And when he was come to the gate of the city, he saw the widow woman gathering sticks, and he called out to her: Give me a little water in a vessel, that I may drink. And when she was going to fetch it, he called after her, saying: Bring me also, I beseech thee, a morsel of bread in thy hand. And she answered: As the Lord thy God liveth, I have no bread, but only a handful of meal in a pot, and a little oil in a cruise; behold I am gathering two sticks, that I may go in and dress it, for me and my son, that we may eat and die. And Elijah said to her: Fear not; but go, and do as thou hast said but make first for me of the same meal a little hearth cake, and bring it to me, and after make for thyself and thy son. For thus saith the Lord, the God of Israel: The pot of meal shall not waste, nor the cruise of oil be diminished, until the day wherein the Lord will

give rain upon the face of the Earth. She went, and did according to the word of Elijah; and he ate, and she, and her house; and from that day, the pot of meal wasted not, and the cruise of oil was not diminished according to the word of the Lord, which he spoke in the hand of Elijah. (I Kings 17:9–16)

Elijah stayed with the family a long time and it happened that the widow's son fell ill and died. The holy prophet prayed grievously to the Lord and then:

He stretched and measured himself upon the child three times, and cried to the Lord, and said: O Lord, my God, let the soul of this child, I beseech thee, return into his body. And the Lord heard the voice of Elijah: and the soul of the child returned into him, and he revived. (I Kings 17:21–22)

This miracle can serve as a source of comfort in our prayer for today's young people who, through the tremendous forces of temptation at work in the contemporary world, are threatened with the loss of their souls. It is also a reminder that there is a restorative power in the depth of prayer and that there are beings such as Elijah of such magnitude as to bring about a regenerative *breath* or pattern of grace.

This healing miracle prefigures a similar healing brought about through Elisha, Elijah's successor, to whom Elijah bequeathed his sacred mantle before he departed from the Earth. The following story of Elisha, serves also as an account of the spiritual rewards of hospitality. Elisha had often been a guest in the Shunamite woman's house. Through the prayers of the holy prophet, the Shunamite woman, who had been barren, became able to conceive and give birth to a son. Like the earlier story of Elijah restoring the life of a child through the urgency of his prayers, Elisha was one day summoned from Mt. Carmel by the Shunamite woman, whose son had fallen ill suddenly and died:

Elisha, therefore, went into the house, and behold the child lay dead on his bed. And going in, he shut the door upon him, and upon the child, and prayed to the Lord. And he went up, and lay upon the child, and put his mouth upon his mouth, and his eyes upon his eyes, and his hands upon his hands; and he bowed himself upon him, and the child's flesh grew warm. Then he returned and walked in the house, once to and fro; and he went up, and lay upon him, and the child gasped seven times, and opened his eyes. (II Kings 4:32–35)

There is yet another prefiguring in this miracle, and that is a direct prophecy of the future New Testament account of the raising of Lazarus from the dead, a miracle in which Elijah himself—in his subsequent incarnation as John the Baptist—became united with Lazarus, as discussed in chapter 12. This miracle is central to the mission of Elijah and is developed further in the unfolding of this story in future chapters. The account of this miracle is a testament to the living waters of life that issue from Jesus Christ. In Elijah's time it was after the miracle of raising a child from the dead had taken place that Elijah was summoned again to confront Ahab and to suggest a test to prove the strength of the God of Israel over Baal and his followers. When this final challenge was met, the rains came:

And Elijah coming to all the people, said: How long do you halt between two sides? If the Lord be God, follow him; but if Baal, then follow him. And the people did not answer him a word. And Elijah said again to the people: I only remain a prophet of the Lord; but the prophets of Baal are four hundred and fifty men. Let two bullocks be given us, and let them choose one bullock for themselves, and cut it to pieces, and lay it on the wood, and put no fire under; and I will dress the other bullock, and lay it upon wood, but put no fire under it. Call ye on the names of your gods, and I will call on the name of my Lord; and the God that shall answer by fire, let him be God. And the people answering said: A very good proposal. (I Kings 18:21–24)

Elijah built an altar of twelve stones, representing the twelve tribes of Israel, and dug a trench around the altar. He prepared the bulls for sacrifice and arranged the wood accordingly:

And he said: Fill four jars with water, and pour it upon the burnt offering, and upon the wood. And again he said: Do it again a second time. And when they had done it the second

time, he said: Do the same also the third time. And they did
so the third time. And the water ran round the altar, and the
trench was filled with water. (I Kings 18: 34, 35)

Although the prophets of Baal cried out to their god and ranted
and raved on until the time of oblation had passed, even cutting themselves with swords until the blood gushed, they received no answer.
But when Elijah prayed, he acknowledged the one God of Israel and
that he, Elijah, was the servant of this God, and that those who had
gone astray needed to bear witness to the Word of God so that their
hearts could be reawakened to their true mission:

Then the fire fell, and consumed the holocaust, and the wood,
and the stones, and the dust, and licked up the water that was
in the trench. (I Kings 18:38)

After this Elijah ordered that all the prophets of Baal, which the Old
Testament counted as four hundred and fifty, be slain. And the people
fell on their faces and believed once more in the one God of Israel.
With this heroic deed, Elijah truly lived up to his name—*Eli* ("my
God") [is] *Jah* (abbreviation for Yahveh, or Yahweh), translated as
"my God is Yahweh."

Then the holy prophet retired to Mt. Carmel to pray for rain. He
requested that his servant go and look toward the sea for the promise
of rain, and then again, to return seven times more, and the seventh
time, Elijah said to his servant:

Behold, a little cloud like a man's hand is rising out of the
sea...and in a little while the heavens grew black with clouds
and wind, and there was a great rain. (I Kings 18:44-45)

This was the third year of the drought and the famine had been severe.
Elijah's choice of action in response to the Word of God—"Go and
show yourself to Ahab; and I will send rain upon the Earth"—had set
things right and the Earth was blessed with God's grace.

Though the people could now eat and drink, Jezebel's fury could
only be quenched with Elijah's death, and so once again Elijah fled a
day's journey into the wilderness, and it was here that Elijah, in the

depths of despair, was touched by an angel and provided with a cake cooked on a hot stone and with a jug of water. Twice, the angel came, each time providing food for Elijah. And this food bore the strength to sustain Elijah for forty days and forty nights during his flight to Mt. Horeb. Elijah dwelt in a cave on Mt. Horeb and called upon the Lord for deliverance, and he witnessed the glory of God in the thunder and lightening and even the shattering of the Earth in an earthquake. Yet the answering Word of the Lord came only in the silence of the moments that followed these awesome events. God's word came in a still small voice. And Elijah was told many things relating to how he was to return and to once again set things right, reestablishing God's kingdom by anointing new leaders over all the nations. And Elijah served toward this mission, but the evil persisted until the last days of Elijah, when he bequeathed his spiritual mantle to Elisha and was taken up into heaven by a chariot of fire:

> Behold a chariot of fire and horses of fire separated the two of them. And Elijah went up by a whirlwind into heaven. (II Kings 2:11)

With this stirring account of Elijah's departure from the Earth, we shall begin our journey, involving an astrological survey into the mysteries inherent in Elijah's role as a guiding force in the evolution of humanity.

3

CONCERNING THE BASICS:
JOHN THE BAPTIST – RAPHAEL

THE BASIC PRINCIPLE OF ASTROLOGICAL REINCARNATION

IN PAYING HOMAGE TO Elijah and the further indications given by Rudolf Steiner of this individuality's incarnations, we have an example of three consecutive incarnations. Before examining this research, let's first consider some basic principles. Reference was made in chapter 1 to the basic principle of comparing the horoscope of birth with the horoscope of death in the preceding incarnation. In the two *Hermetic Astrology* volumes, this rule is extended to include the comparison of the horoscope of conception (also referred to as the epoch) with the birth horoscope of the previous incarnation. If we juxtapose these in a sequence of three incarnations: A, B, and C, a pattern emerges, expressed in figure 2, with the following symbols: Ø = conception/epoch, ✳ = birth, † = death.

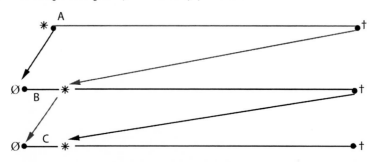

Figure 2: A pattern extending over three consecutive incarnations

Here we see a line connecting †A to ✳B to ØC, showing a pattern (in terms of astrological reincarnation research) between death (†) in one incarnation to birth (✳) in the following incarnation, and, further, to conception (Ø) in the next incarnation. In figure 2, the pattern extends over three *consecutive* incarnations, signifying that it is meaningful to do chart comparisons over several incarnations. It has been found, however, that even if the incarnations are not consecutive it is possible to do chart comparisons not only with the birth and death horoscopes of the previous known incarnation, but also with the birth and death horoscopes of the incarnation prior to that incarnation and extending further back to still earlier incarnations.

The astrological reincarnation research developed initially in the two *Hermetic Astrology* volumes and demonstrated further in this book is a fruit of the seeds sown by Rudolf Steiner. As a beginning point of reference in chapter 1, Rudolf Steiner is quoted with respect to the basic principle of comparing the birth horoscope with the death horoscope in the preceding incarnation.

One of Rudolf Steiner's close collaborators, the scientist Guenther Wachsmuth, did some research into this, which was published in 1956 in his book *Kosmische Aspekte von Geburt und Tod* ("Cosmic Aspects of Birth and Death"). Prior to Guenther Wachsmuth's investigation, mathematician and astronomer Elisabeth Vreede—who was another of Rudolf Steiner's close collaborators—encouraged Willi Sucher, my teacher and the pioneer of astrosophy, to carry forward with Rudolf Steiner's indications for a new astrology. Throughout his long life of research into star wisdom, which began in the 1930s and lasted until his death in 1985, Willi Sucher returned over and again to consider the birth horoscope in relation to the preceding incarnation's death horoscope. Although he made no systematic presentation, his research findings can be found scattered through various books and journals.[30]

It was Willi Sucher's early pioneering work looking for the significance of the life of Christ in relation to the wisdom of the stars that was foundational in my empirical research regarding the *dating of the Christ events* in relation to the stars.[31] It is this earlier research into the events of the life of Christ that has made it possible to include the life of John the Baptist (as revealed by the stars) in connection with the series of incarnations related to Elijah.

Elisabeth Vreede, Guenther Wachsmuth, and Willi Sucher all used the astronomical zodiac.[32] The astronomical zodiac is not divided into degrees and minutes, so that the chart comparisons drawn in terms of the astronomical zodiac can only be approximate. My research follows in the tradition of Rudolf Steiner, Elisabeth Vreede, Guenther Wachsmuth, and Willi Sucher. However, through my research I have discovered that it is possible to make extremely accurate comparisons between the birth horoscope of one incarnation and the death horoscope of the preceding incarnation by using the sidereal zodiac. Because the sidereal zodiac is divided into degrees and minutes in relation to the fixed stars, each planet's position is given accurately in relation to the stars.[33] The significance of this can best be illustrated by the following example:

John the Baptist-Raphael

As mentioned in the introduction, Jesus Christ revealed to his disciples that Elijah reincarnated as John the Baptist. According to Rudolf Steiner, the individuality of John the Baptist then reincarnated as the Renaissance painter Raphael, who was born on March 28, 1483 (Good Friday). Can this be proven empirically through astrological research? And what can be learned through such an enquiry?

As shown in *Chronicle of the Living Christ*,[34] John the Baptist was born on June 4 in the year 2 B.C.E.[35] Drawing upon this line of research, we discover that the birth horoscope of John the Baptist[36] reveals that the planet Pluto (geocentric) was in the 23rd degree (22 degrees 44 minutes) of sidereal Virgo that, in astrological terms, is written 22♍44 (see page 22). Keeping in mind that the planets in the horoscopes from one incarnation to later incarnations tend to reveal patterns of alignment in the sidereal zodiac, we see a correspondence at Raphael's birth when again geocentric Pluto was in the twenty-third degree of Virgo (22♍20), just twenty-four minutes (approximately one-third of a degree) from its position at John the Baptist's birth. Here we can speak of an *alignment*—in astrological terms a conjunction—in the 23rd degree of sidereal Virgo in the birth horoscopes of the two incarnations. As this alignment takes place on the same side of the zodiac it is written "conjunct," whereas if there is an alignment of planets on opposite sides of the zodiac, it is written "opposite."

The above example illustrates the relevance of the sidereal zodiac to astrological reincarnation research. This becomes transparently clear when we consider the corresponding positions of Pluto in the tropical zodiac: indicating that Pluto was in Virgo at 19♍40 in the tropical zodiac at the birth of John the Baptist and in Libra at 9♎52 in the tropical zodiac at the birth of Raphael (see page 24). The Pluto positions in the tropical zodiac are nowhere near an *alignment* or *conjunction*, since the two positions are over twenty degrees apart.

This discrepancy of over 20° between the positions of Pluto in the tropical horoscopes is due to the retrograde movement through the astronomical constellations of the vernal point (defining the zero degree of the tropical zodiac) at a rate of 1° every seventy-two years.[37] Thus we have a clear demonstration of how misleading the use of the tropical system can be when comparing horoscopes from one incarnation to the next, since the planetary positions between the incarnations being considered, because their true astronomical locations are no longer depicted, do not reveal the planetary alignments from one incarnation to the next. In contrast, the sidereal zodiac is fixed, defined in relation to the fixed stars, and therefore the sidereal horoscopes depict the actual *astronomical* locations of the planets. This example shows what a powerful tool the sidereal zodiac—originally defined by the Babylonians—is for astrological reincarnation research.

Through the findings of more than thirty years of astrological reincarnation research, there is now empirical proof that the sidereal zodiac is valid as an accurate indicator or frame of reference for the reincarnating soul. What can we discern from this research?

The results of research into the reincarnations indicated by Rudolf Steiner have verified that from one incarnation to a later incarnation the planets tend to align either in conjunction at the same location in the sidereal zodiac or in opposition at opposite sides of the sidereal zodiac. It was in analyzing the results of the research itself that it emerged that these alignments were only to be found in the sidereal zodiac and not in the tropical zodiac, as the example John the Baptist–

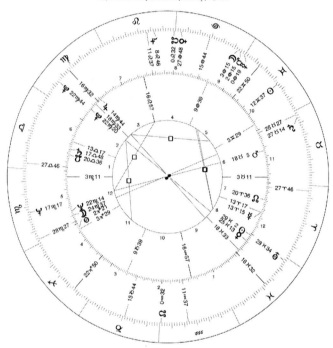

Comparison Chart

Outer - Geocentric
Birth of John the Baptist
At Jutta (Yattah), Latitude 31N32', Longitude 35E6'
Date: Wednesday, 4/JUN/2 B.C., Julian
Time: 0:37, Local Time
Sidereal Time 17:15:23, Vernal Point 3♈ 3'52"

Inner - Geocentric
Birth of Raphael
At Urbino/Italy, Latitude 43N44', Longitude 12E38'
Date: Friday, 28/MAR/1483, Julian
Time: 21:27, Local Time
Sidereal Time 10:25:25, Vernal Point 12 ♓27'56"

House System: Placidus, Zodiac: Sidereal SVP
Aspect set: Conjunction/Square/Opposition

Geocentric Sidereal Alignment Example:
John the Baptist Birth g-Pluto at 22♍44—conjunct—
Raphael Birth g-Pluto at 22♍20

Raphael shows. On this account I discontinued using the tropical zodiac in the early 1970s.

Drawing upon the sidereal zodiac as a *reliable* background for the movement of the planets and thus *foundational* for astrological reincarnation research, what can we now conclude from the reincarnation example of John the Baptist–Raphael?

Assuming that Rudolf Steiner's reincarnation indication that John the Baptist reincarnated as Raphael is true, then the comparison of the birth chart with the birth chart of the former incarnation tells us that the soul of Raphael chose to incarnate at the completion of six orbits of Pluto around the sidereal zodiac following his incarnation as John the Baptist. I say six orbits because one orbit takes about 247.3 years, and multiplying this by six amounts to 1483.8 years, which is approximately the time between John the Baptist's birth in 2 B.C.E. [astronomically 2 B.C.E. equals the year -1][38] and Raphael's birth in 1483. The exact time intervening between the two births is 1483.813 years.

GEOCENTRIC AND HELIOCENTRIC / HERMETIC PERSPECTIVES[39]

The implication here is that between incarnations, the human soul perceives the movement of the planets against the background of the fixed stars of the sidereal zodiac. According to the astrological reincarnation research presented in this book, it is possible to show that this perception is both geocentric *and* heliocentric/hermetic. For example, the second "law" of astrological reincarnation states that alignments also occur from death in one incarnation to birth in the next incarnation in terms of the *heliocentric* positions of the planets (in the sidereal zodiac). This takes into account the vantage point of the Sun, where the heliocentric positions of the planets are included in the hermetic horoscope.

In the example we have just considered, the Pluto alignment is in terms of the *geocentric* positions in the sidereal zodiac. The corresponding *heliocentric* positions of Pluto also reveal an *alignment* (although not so exact): heliocentric Pluto was in sidereal Virgo at 24♍43 at the birth of John the Baptist and 22♍34 at the birth of Raphael (about two degrees apart; see appendix 2, chart 5). Whereas *geocentrically* they are in *close conjunction* (about one-third of a degree apart), *heliocentrically* they are about two degrees apart.

In the interest of a more expanded perspective, this book will include examples of both geocentric and heliocentric alignments against the background of the sidereal zodiac.

What are the implications for reincarnation research following upon the results of finding planetary alignments in both geocentric *and* heliocentric/hermetic horoscopes?

Comparison Chart

Outer - Geocentric	Inner - Geocentric
Birth of John the Baptist	Birth of Raphael
At Jutta (Yattah), Latitude 31N32', Longitude 35E6'	At Urbino/Italy, Latitude 43N44', Longitude 12E38'
Date: Wednesday, 4/JUN/2 B.C., Julian	Date: Friday, 28/MAR/1483, Julian
Time: 0:37, Local Time	Time: 21:27, Local Time
Sidereal Time 17:15:23	Sidereal Time 10:25:25

House System: Placidus, Zodiac: Tropical
Aspect set: Conjunction/Square/Opposition

Geocentric Tropical Comparison: John the Baptist Birth g-Pluto
at 19♍40—20° apart—Raphael Birth g-Pluto at 9♎52

The Hermetic (Tychonic) Perspective[40]

A remarkable chapter in history was given birth through the meeting of the Danish astronomer Tycho Brahe (1546–1601) and the German astronomer Johannes Kepler (1571–1630). Their association lasted some 20 months, until Tycho's death on October 24, 1601. On his death-bed Tycho Brahe was heard to repeat over and over again the words, "Let me not seem to have lived in vain." Tycho was requesting that the hermetic/Tychonic astronomical perspective be understood by the world as important. At that time the heliocentric perspective was emerging as the rival of the traditional geocentric view, due to Copernicus' discovery that the planets revolve around the Sun.[41] Tycho Brahe knew intuitively that, even though the planets revolve around the Sun, the Earth can still be considered as the center. Although not astronomically viable, as it turns out, his system is *astrologically* true, as will emerge in the course of this book.

The heliocentric/hermetic horoscope brings to fulfillment the fervent request and hope of Tycho Brahe that the Tychonic astronomical system not be forgotten. The term "heliocentric/ hermetic" signifies a heliocentric (Sun-centered) perspective that *includes* the geocentric (Earth-centered) location of the Sun, Moon, and Moon's Nodes, as well as the Ascendant/Descendant, Midheaven, and other key astrological house indicators—all of which are otherwise omitted from the purely heliocentric perspective. Thirty years of empirical reincarnation (karma) research indicate that by combining *both* perspectives *hidden* mysteries concerning reincarnation and destiny are revealed.

Going a Step Further

Including both the geocentric and the heliocentric perspectives— the latter in the hermetic chart—takes into account the two levels of perception of the reincarnating soul.[42] One level of perception has to do with the higher self of the human being—taking into consideration that the true home of the higher self is the Sun. This is the perspective referred to by Novalis in reference to *lifting the veil of Isis* and discovering *the Self.* This is the Self *unveiled* as an emanation from the Sun. The other level has to do with the human "I" that is connected with the Earth. The human "I" reincarnates upon the Earth and thus has a *geocentric* perspective (geocentric signifies "Earth centered," where *geo = Ge = Gaia,* the Earth goddess). On the other hand, the "I" of the higher self indwells the Sun and has a *heliocentric* perspective (heliocentric means "Sun centered," derived from the Greek word *Helios* for the Sun god). These two vantage points of the "I"—the Sun and the Earth—can be conceived of as "eyes" placed vertically one above the other. With our normal sense of sight, the two eyes are placed horizontally alongside one another, each with its own perspective. However, our ability to see requires that the two eyes also "see together." Correspondingly the higher "I" (Sun) and the lower "I" (Earth) also have their individual perspectives, and yet one of the most important goals of spiritual development is that they learn to "see together" vertically.

For this reason both charts—geocentric and heliocentric/hermetic[43]—are relevant and need to be seen together. This is how, generally speaking, one can come to a comparative analysis of an individual's horoscope.

Comparative Chart Analysis

When I say "comparative analysis," I hasten to add that the form used in this presentation is entirely due to the Astrofire computer program developed by astrosopher Peter Treadgold.[44] All the horoscopes presented in this book have been computed using the Astrofire program. The Astrofire program makes it easy to compare any two geocentric or any two hermetic horoscopes. In the reincarnation examples presented in this book, with both geocentric and hermetic chart comparisons, one horoscope is presented in the center and the other horoscope is presented around the outside. Using such comparison horoscopes, it is therefore possible to see planetary alignments at a glance.

It is thanks to the work of Peter Treadgold that astrological reincarnation research—on the scale undertaken in this book—has been made possible. Just imagine that Elisabeth Vreede, Guenther Wachmuth and Willi Sucher all made their computations by hand! Peter's contribution to the ongoing research of astrosophy has made possible a ground-breaking step toward the birth of a new astrology for the new millennium—an approach to astrology based on a conscious knowledge of the basic principles and "laws" of astrology discovered through astrological reincarnation research.

THE CONCEPTION (EPOCH) HOROSCOPE

The "hermetic rule" for determining the epoch:

The Moon's location at conception becomes the place of the Ascendant at birth; similarly the Ascendant at conception becomes the location of the Moon at birth. Given a birth horoscope based on a reasonably accurate time of birth, using the Astrofire program it is possible to compute the conception (epoch) horoscope.[45] Since the astrological "law" (hermetic rule)[46] for computing the epoch is open-ended, several possibilities are given for the epoch and from these, one epoch must to be chosen.

Assuming that it is possible to find the correct epoch horoscope, each historical personality will have horoscopes cast for three significant moments: epoch, birth, and death. If a reincarnation is also in question, then there are additional horoscopes relevant to the research, as may be seen from figure 2 at the start of this chapter. For a thorough investigation, in looking for new "laws" of astrological reincarnation, each of the horoscopes from the one incarnation should be compared with each of the horoscopes from the other incarnation cast for epoch, birth and death. This means nine chart comparisons for a reincarnation analysis. What happens if three successive incarnations are known, each with their corresponding dates?

If the birth time is known in each incarnation, then it is possible to compute the epoch using the hermetic rule, so that each incarnation has relevant horoscopes for incarnations A, B, and C (see figure 2). Then there are nine chart comparisons between A and B, nine between A and C, and nine between B and C, making twenty-seven chart comparisons in all. Generally speaking, however, the birth time of historical personalities is not known, particularly the further one goes back in time, in which case there is no way of determining the epoch. This immediately reduces the number of chart comparisons as will be evident in the example of the three consecutive incarnations that we shall now consider.

Before we begin our investigation, let us keep in mind that in seeking to understand the analysis, we are searching for the true *foundations* of astrology. We are looking to uncover "nuggets of gold," informing us about the "why" and "how" of a horoscope. Why do the planetary alignments appear in particular patterns in the sidereal zodiac in the horoscopes between epoch, birth and death of reincarnating individualities?

The Akasha Chronicle

Let us first consider the time of death. What takes place when the physical body is laid aside at the moment of death? At death, no longer attached to the physical body, the human "I" indwells the etheric and astral bodies. [47] With the release of the etheric or life body from the physical body at the moment of death, two events occur simultaneously—the etheric life body departs, leaving behind a corpse, while at the same time the "I" experiences the content of the etheric body such that images of biographical events appear as a great panorama or flashback of the memory content of that lifetime. Many people who have undergone "near death experiences" have reported the panoramic flashback of their life. This is the subjective experience. Now let us consider what is taking place objectively.

Just as the physical body is laid aside at the moment of death, there begins now the laying aside of the etheric body, usually taking about three days, which is called the "second death." The physical body decomposes—"ashes to ashes, dust to dust"[48]—while at the same time the etheric body is re-membered back into the planetary (etheric) cosmos. During life the etheric body not only maintains all the life functions of the physical body (breathing, circulation, etc.), but also acts as a repository for all memories. Thus our memories are imprinted into our etheric body. When the etheric body is re-membered back into the etheric cosmos during the three days or so following death, the biographical memories of that incarnation become imprinted into the etheric cosmos, which extends throughout the planetary spheres up to the realm of the fixed stars. What does this signify for our memories?

The panoramic flashback at the moment of death is actually a biographical summary of the life or incarnation completed at that moment, which becomes offered up through the planets to become imprinted into the constellations of the fixed stars where the planets are located at the moment of death. Seen in this light the sidereal zodiac represents the outer aspect of what is known as the "akasha chronicle" containing the memory of everything that has ever happened, including all the events of our lives. The planetary configuration at the moment of death is thus highly significant as an expression of the summary of the destiny lived through in that incarnation. Correspondingly, the planetary positions indicated in the death horoscope, when viewed against the background of the sidereal zodiac, reveal the zodiacal locations of where our memories are imprinted into the akasha chronicle within the etheric life body of the cosmos.

According to astronomical reincarnation research it appears that the human being—when the time comes to reincarnate—chooses to be born at a moment when significant planets align with their sidereal positions at death in the preceding incarnation (both geocentric and heliocentric, taking account of the perspective of the earthly personality and also of the solar higher self). In this way, through the planetary positions indicated in the sidereal zodiac, one can discover how memory is carried over from one incarnation to the next.

Picking up the Thread

In comparing birth and death horoscopes one can find planetary alignments on the same or on opposite sides of the sidereal zodiac, revealing that the cosmic memory is activated both when a planet is in opposition or in conjunction with its position at a prior or subsequent event (epoch, birth, death).

These patterns of alignment reveal the different perspectives of the reincarnating soul. We can think of conjunction as a direct activation of the karmic memory and opposition as an indirect activation of a karmic memory. The opposition may indicate the potential for a "full flowering" or completion of a former memory, aptitude, or event. Thus, if the karmic memory is associated with a particular star with which a planet is in conjunction, a "*geocentric* conjunction" alignment (*Earth*—Planet—Star), indicates a different perspective from a "*heliocentric* conjunction" alignment (*Sun*—Planet—Star). Whereas with a "*geocentric* opposition" the alignment is (Planet—*Earth*—Star), and a "*heliocentric* opposition" is: (Planet—*Sun*—Star). In all cases it is a matter of alignment. In the case of conjunction the planet activates the karmic memory stored in the star directly, and in the case of opposition the planet activates the memory indirectly via the Earth (geocentrically) or the Sun (heliocentrically).

This book will refer to alignments using the word *conjunct* to indicate a conjunction alignment and *opposite* to signify an opposition alignment.

Finding the Epoch (Conception)

The foregoing is clearly relevant for the connection between the death horoscope and the birth horoscope of the next incarnation. However, what does this signify for the *epoch*? We shall follow the obvious lead that the epoch/conception configuration is connected to the birth horoscope (occurring in most cases approximately nine months before birth). However, because we are working with a reincarnation example, we shall refer to the *birth/death* horoscopes of the *previous* incarnation.

In choosing a birth horoscope to *resonate* with the preceding death horoscope, the reincarnating soul must necessarily also choose an epoch/conception horoscope. As illustrated in figure 2, we see that the gaze of the reincarnating soul, in making the choice of the epoch/conception configuration, is directed back to the birth horoscope of the preceding incarnation.

In order to gain a complete picture (indicated by figure 2) we can imagine that the gaze of the reincarnating soul is directed back to the destiny lived out in the former incarnation, which began at birth (and conception) and extended until death. Thus the reincarnating soul beholds the birth and conception horoscopes of the preceding incarnation as the "door" through which the human soul entered into that incarnation, and seeks to find a connection with this by choosing an epoch/conception configuration for the approaching new incarnation that *resonates*—in terms of planetary positions in the sidereal zodiac—with planetary positions as they had been at the *opening of the door* to the preceding incarnation.

At the same time the reincarnating soul beholds the "exit" from the preceding incarnation—which has been *re-membered* by the planetary configuration at death—and seeks to harmonize this with the birth configuration for the approaching new incarnation.

The foregoing can be regarded as a "working hypothesis" for our investigation into three consecutive incarnations. For the sake of objectivity, however, we must forego any prior assumptions (concerning special connections between the death horoscope and the subsequent birth horoscope). Rather, we shall undertake all possible comparisons—geocentric and heliocentric/hermetic—between the various epoch, birth, and death horoscopes.

In making these comparisons we are searching for the ultimate "laws" and principles of astrology. We are embarking upon empirical research looking for objective facts, and we shall see what we discover. Would it not be extraordinary, for example, if we were to find that the position of the Sun chosen by the reincarnating soul was related to the position of Jupiter in the last-but-one incarnation? Obviously, in making so many chart comparisons, we shall look for only the most striking alignments, as in the example of the geocentric alignment of Pluto in the birth horoscope of John the Baptist to the birth horoscope of Raphael. And, as we shall see, another outer planet, Uranus, is involved in the alignment that took place between the death of Raphael and his rebirth 252 years later as Novalis (see appendix 2, charts 10 & 11).

4

NOVALIS

AFTER THE SEQUENCE OF incarnations: Elijah—John the Baptist—Raphael, Rudolf Steiner indicates that Raphael later reincarnated as the German Romantic poet Novalis (1772–1801).

In the interest of a backward gaze, in contemplating the series of incarnations relating to Elijah, we shall begin our investigation with the Novalis individuality. Just as the reincarnating soul's gaze is directed back to the destiny lived out in its former incarnations, so too we shall contemplate the *resonating* aspects of the Novalis horoscopes as a "looking glass" to his preceding incarnations in order to discover the "golden thread" of connection. Our purpose is to reveal the primary focus of this individuality's mission from life to life.

Novalis was born in Wiederstedt, Germany, on May 2, 1772, at about ten A.M. He died at Weissenfels, Germany, on March 25, 1801, at about 12:30 P.M. Since we know his time of birth, we can calculate the epoch/conception, applying the hermetic rule. One possibility for the epoch is August 9, 1771, 9:53 P.M. *local time.*[49]

Our first task is to investigate the plausibility of this epoch/conception configuration by comparing it with the death horoscope of Novalis. The reason for doing this is that the epoch signifies the beginning of the *formation* of the etheric body from the etheric cosmos, and the death horoscope shows the beginning of the *re-membering* of the etheric body back into the etheric cosmos.

Novalis: Conception–Death Horoscope (Geocentric)

In the geocentric comparison of the epoch/conception chart of Novalis with his death chart, we find an *alignment* of the Moon in sidereal Cancer: *Epoch Moon 14♋19/Death Moon 16♋25.*[50] The

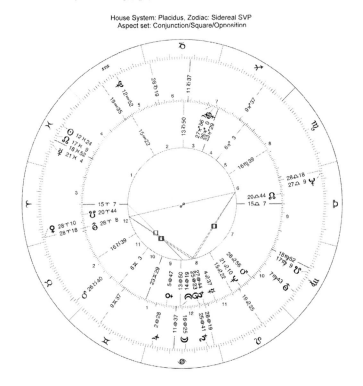

Comparison Chart

Outer - Geocentric
Death of Novalis
At Weissenfels, Germany, Latitude 51N12', Longitude 11E58'
Date: Wednesday, 25/MAR/1801, Gregorian
Time: 12:30, Local Time
Sidereal Time 24:39:48, Vernal Point 8 ℋ 2' 1"

Inner - Geocentric
Epoch of Novalis
At Wiederstedt, Latitude 51N38', Longitude 11E30'
Date: Friday, 9/AUG/1771, Gregorian
Time: 21: 7, Time Zone GMT
Sidereal Time 19: 5:17, Vernal Point 8 ℋ 26'49"

House System: Placidus, Zodiac: Sidereal SVP
Aspect set: Conjunction/Square/Opposition

Epoch g-Uranus 28♈08—aligns—Death g-Venus 28♈10

Moon is thought to be the repository of our *soul karma* and the whole sphere bounded by the Moon's orbit around the Earth is traditionally considered to be the domain of our guardian angel. Near the center of Cancer, at 12½°♋, there is a beautiful cluster of stars called *Praesepe,* which was revered by ancient astrologers as a most potent location in the zodiac.[51] This cluster of stars, also known as the "Beehive," occupies a central position in the sidereal sign of Cancer. This sign was believed to be the place where cosmic souls enter into earthly life in order to gain wisdom that they carry back to their "queen" after death (the queen of heaven, known in the Gnostic tradition as Sophia). This picture that resounds in the Moon's location in the conception and death horoscopes of Novalis strikes a resonant chord with the poetry of Novalis—whose chosen pen name *Novalis* means "entering into new terrain" in Latin. Novalis's poetry is permeated with the mood of Sophia.

This theme is echoed once again with the geocentric alignment of Uranus and Venus in sidereal Aries. *Cassiopeia*, the constellation of the heavenly queen, stretches her starry body above this region of the heavens. Looking at the geocentric sidereal comparison chart of Novalis's conception/epoch with his death configuration (on the previous page), most striking in this comparison chart is the discovery of the alignment of the *Sun* and *Saturn* in Cancer in the conception/epoch and death horoscopes of Novalis: *Epoch Sun 25♋24/Death g-Saturn 25♋41.*[52] In other words at the moment of Novalis's death, Saturn (which is the repository of our *spiritual karma*) was transiting the position of the Sun at his conception (epoch). The Sun, signifying the source of life, at Novalis's conception/epoch (indicating the beginning of the *weaving* of the "life principle" into the etheric body) was transited by Saturn, the planet traditionally associated with *recalling* the soul back to the spiritual world at the time of death. The symbolism here speaks strongly in favor of our having found the correct epoch for Novalis on the evening of August 9, 1771.

The Mystery of Inspiration: Novalis—Mary of Nazareth

There is yet another reason for regarding the Sun's position in sidereal Cancer (25♋24) at epoch as authentic, for this coincides exactly with the position in sidereal Cancer (25♋25) of the Sun at the birth of Mary of Nazareth.[53] Novalis was inspired by Mary, as his poetry reveals:

> Who once have seen thee, Mother fair,
> Is saved from ruin and despair;...
> Often in dreams I see thee clear
> Before my inmost soul appear.
> The Child-God from thy arm is bending....
> Now glorified, O Queen divine,
> Accept from me my heart and life as Thine.
> Thou knowest well, beloved Queen,
> That I completely Thine have been.[54]

One can feel in the poet's devotion to Mary, the stream of inspiration remembered from one incarnation to the next. From Raphael's beautiful series of paintings of the Madonna and Child to the poetry of Novalis, there is a metamorphosis from image into the *living* word.

In countless pictures I behold Thee
Maria, clothed in form divine;
But none so wondrously enfold Thee
As in my soul I see Thee shine.

I only know the world is flowing
Away from me like restless dreams;
A Heaven more sweet than mortal knowing
Within my soul forever gleams.[55]

In this relationship between the Sun at Novalis's conception and the Sun at the birth of Mary of Nazareth (also known as the Nathan Mary), we touch upon the mystery of *inspiration*. The gaze of the reincarnating soul looks back not only to its own previous incarnations, but also to the incarnations that represent a source of inspiration. Here again astrological reincarnation research based on the sidereal zodiac reveals itself to be a powerful tool, making it possible to uncover and reveal streams of inspiration—as in this example, with Novalis clearly being inspired by Mary (see following page for the comparison chart).

In the hermetic epoch/conception chart of Novalis, the position of *Venus* in sidereal Gemini is significant: (7Ⅱ36). Let us recall that Novalis had an earlier incarnation as John the Baptist, whose greatest deed in that incarnation was the baptism of Jesus in the River Jordan. In the hermetic chart of the baptism in the Jordan, Mercury (7Ⅱ45) and Saturn (7Ⅱ18) are in conjunction in sidereal Gemini.[56] Here we discover that the heliocentric/hermetic position of Venus in Gemini at Novalis's epoch thus recalled the memory of the Baptism in the Jordan, which was magnified by the heliocentric hermetic *conjunction* of *Mercury* and *Saturn (remembered* in sidereal Gemini), and inscribed into the akasha chronicle as one of the great events in world history—in which Novalis in his incarnation as John the Baptist had played the significant role of baptizer for Jesus (see page 37 for the comparison chart).

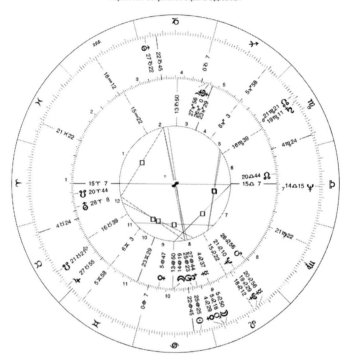

Comparison Chart

Outer - Geocentric	Inner - Geocentric
Birth of Mary Nathan	Epoch of Novalis
At Nazareth, Latitude 32N42', Longitude 35E18'	At Wiederstedt, Latitude 51N38', Longitude 11E30'
Date: Sunday, 17/JUL/17 B.C., Julian	Date: Friday, 9/AUG/1771, Gregorian
Time: 23:51, Local Time	Time: 21: 7, Time Zone GMT
Sidereal Time 19:24:27, Vernal Point 3♈16'13"	Sidereal Time 19: 5:17, Vernal Point 8♓26'49"

House System: Placidus, Zodiac: Sidereal SVP
Aspect set: Conjunction/Square/Opposition

*Novalis's conception/epoch chart (inner circle) compared with the
birth chart of the Nathan Mary (outer circle): alignment of the Sun at
Novalis's conception with the Sun at the birth of the Nathan Mary.*

Here again astrological reincarnation research reveals itself to
be a very precise tool in uncovering the mysteries of destiny. At the
same time this example shows the enormous significance of the epoch
horoscope. It is to the credit of Willi Sucher that he realized that the
epoch configuration is significant in its own right, and not simply as
a step in the application of the hermetic rule to rectify the time of
birth, which in the astrological world is generally the sole use of the
determination of the epoch.[57]

Comparison Chart

Outer - Tychonic	Inner - Tychonic
Baptism of Jesus Christ	Epoch of Novalis
At River Jordon, near Ono, Latitude 31N50', Longitude 35E33'	At Wiederstedt, Latitude 51N38', Longitude 11E30'
Date: Friday, 23/SEP/29, Julian	Date: Friday, 9/AUG/1771, Gregorian
Time: 10: 0, Local Time	Time: 21: 7, Time Zone GMT
Sidereal Time 10: 0: 1, Vernal Point 2♈38'42"	Sidereal Time 19: 5:17, Vernal Point 8♓26'49"

House System: Placidus, Zodiac: Sidereal SVP
Aspect set: Conjunction/Square/Opposition

Novalis's hermetic epoch chart reveals an otherwise "hidden"
resonance with the memory of the Baptism in the Jordan. In the
Astrofire program used to calculate the horoscopes presented in this
book, the hermetic horoscope is referred to as the Tychonic chart in
honor of Tycho Brahe, since the hermetic horoscope is an application
of the Tychonic system that he discovered in 1583.

Novalis: Birth and Death (Geocentric)

In determining Novalis's epoch through applying the hermetic
rule (sometimes called the "Trutine of Hermes"), the time of Novalis's
birth is now rectified to 10:08 A.M. local time. In casting a glance
at the *geocentric* comparison between Novalis's birth and death
horoscopes, we see that there was an *alignment of the Sun in Pisces*
(12♓24) at death with g-Mars (13♓20) at birth. This alignment is

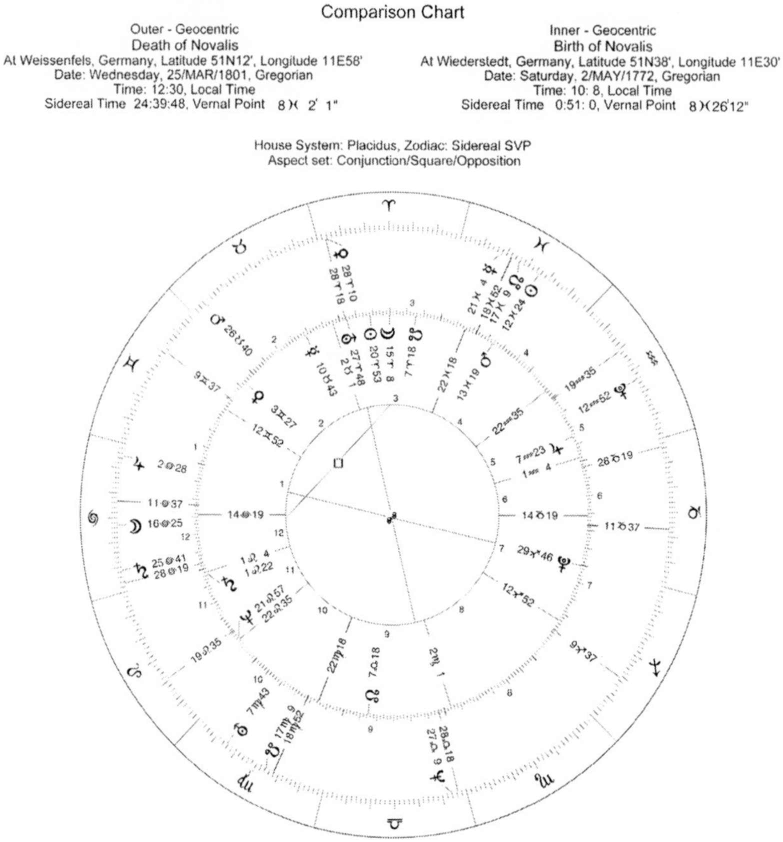
Comparison Chart
Outer - Geocentric
Death of Novalis
At Weissenfels, Germany, Latitude 51N12', Longitude 11E58'
Date: Wednesday, 25/MAR/1801, Gregorian
Time: 12:30, Local Time
Sidereal Time 24:39:48, Vernal Point 8♓ 2' 1"
Inner - Geocentric
Birth of Novalis
At Wiederstedt, Germany, Latitude 51N38', Longitude 11E30'
Date: Saturday, 2/MAY/1772, Gregorian
Time: 10: 8, Local Time
Sidereal Time 0:51: 0, Vernal Point 8♓26'12"
House System: Placidus, Zodiac: Sidereal SVP
Aspect set: Conjunction/Square/Opposition

within one degree and is therefore noteworthy. It marks Novalis's sudden and unexpected death at the age of 28 through tuberculosis. Using Novalis's "magic wand of analogy"[58] we have a striking image here of Pisces, the two fishes: one swimming toward the constellation of Aquarius, representing the future, connected by a *cord* to the second fish, which is swimming upward toward heaven.

Also note that *Mars in Pisces (13♓20) at Novalis's birth is close to the Midheaven in Pisces (22♓18)*, representing the zodiacal high point in his geocentric birth horoscope. The placement of a planet close to the Ascendant or to the Midheaven often indicates that it is of primary significance in the person's life.[59] The planet Mars, therefore, was extremely important in the life of Novalis, albeit the higher aspect of Mars, which is concerned with the word, since the Mars center in the human being is located in the region of the larynx.[60]

Given also the higher aspect of Pisces—the profound moral impulse communicated by Christ[61]—Mars in Pisces at Novalis's birth indicates a coming into manifestation on the Earth of the "moral word," this being the primary concern in Novalis's writings, whereby the *pull of the cord of connection* to heaven is indicated by the Sun in Pisces at Novalis's death. Central to Novalis's life was the "magic word"[62]— the highest expression of Mars.

The comparison of the birth horoscope with the death horoscope *in the same incarnation* is not complicated to do, as it does not require the application of the hermetic rule. It does, however, often reveal striking alignments. It is a kind of "preliminary exercise" prior to making such comparisons in successive incarnations. Quite often it reveals the extraordinary significance of the heliocentric/hermetic horoscope, in which alignments frequently occur between the birth and death horoscopes. In this way one can discover for oneself the astrological validity of the heliocentric/hermetic horoscope, without taking recourse to the more complex level of astrological reincarnation research.

Novalis: Death/Resurrection—Initiation/Baptism

Novalis died at Easter time, March 25, 1801. According to accounts from his circle of friends, the period of time leading up to Easter—referred to by Novalis as "silent week"—constituted a "constellation of time" that was of monumental significance in Novalis's biography. March 15, 1795, marked the date of Novalis's engagement to his beloved Sophie von Kühn; March 17, 1782, was Sophie's birth date; and March 19, 1797, was the day of her death. About one month later, his favorite brother Erasmus died on Good Friday, April 14, 1797.

The words engraved on the inside of Novalis's engagement ring, "Sophia be my guardian spirit,"[63] proved to be prophetic for Novalis, when on May 13, 1797, at Sophie's gravesite (in Thüringen, Germany), Novalis experienced a spiritual awakening, a *resurrection* in the sense of breaking through, from an intense and even life-threatening period of grief (following Sophie's death), to an experience of a dying *into* birth and a reconnection with the starry heavens.

Comparison Chart

Outer - Tychonic
Gravesite Initiation of Novalis

Inner - Tychonic
Birth of Novalis
At Wiederstedt, Germany, Latitude 51N38', Longitude 11E30'

Date: Saturday, 13/MAY/1797, Gregorian
Time: 20: 0, Local Time
Sidereal Time 11:28: 3, Vernal Point 8 ♓ 5'15"

Date: Saturday, 2/MAY/1772, Gregorian
Time: 10: 8, Local Time
Sidereal Time 0:51: 0, Vernal Point 8 ♓ 26'12"

House System: Placidus, Zodiac: Sidereal SVP
Aspect set: Conjunction/Square/Opposition

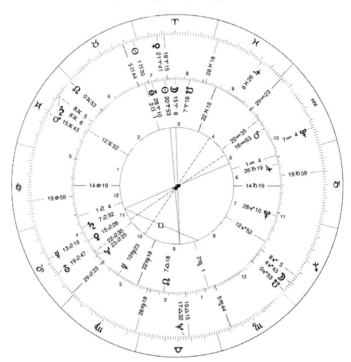

This was to be the cosmic origin of his creative genius as a poet and prophetic spokesman for the future—calling for an all-embracing Sophianic Christianity, born of freedom and lived as an expression of the "wholeness of the human being." This birth was a *death* in the real sense of the smaller "I" dying into the birth of his larger destiny, a birth into the true significance of his higher "I," as a dynamic expression of his *eternal* sense of "I"-ness.

Referring now to this miracle—Novalis's visionary experience at the gravesite of Sophie von Kühn, as an *initiation and baptism*—let us compare the heliocentric/hermetic horoscope of the day of the vision, May 13, 1797, with Novalis's hermetic birth horoscope in order to discover the *hidden* stellar implications of this day, as it relates to the destiny of Novalis (see horoscope above).

We can observe in the comparison of the two horoscopes that this is a "Venus mystery," with the planet *Venus (21♈41) transiting the place of the Sun's location in Aries (20♈53)* at the time of Novalis's birth, noting that—since this is the hermetic chart comparison—it is *heliocentric* Venus that is transiting his birth Sun. The Sun at Novalis's birth, representing the perspective of the higher self, or eternal "I," was transited by *heliocentric* Venus at this event of his spiritual awakening as a poet of extraordinary genius. Here the cosmic symbolism is clear: Venus, representing his beloved Sophie, played the role of muse to his "I"—not, however, geocentric (earthly) Venus, but heliocentric (heavenly) Venus, since she approached him from heavenly realms.

In this hermetic chart comparison (see previous page), bearing in mind that the human heart corresponds to the sign of Leo, it is also interesting to observe that heliocentric Mercury in Leo (13♌18) is transiting the location of heliocentric Venus (15♌28), in turn in opposition to heliocentric Mars (16♒53) when Novalis was born (the significance of the heliocentric opposition between Venus and Mars in Novalis's hermetic birth chart will be considered later). The transit of heliocentric Mercury, seen as an agent of alchemy and healing, over heliocentric Venus in Novalis's birth horoscope finds its reflection in the prophetic words written by Novalis ten days following Sophie's death: "Sophia restored the lost throne to my heart."[64]

In the fourth of his *Sacred Songs*, Novalis wrote of the experience:

> While I thus in sickness languished,
> All my heart with sorrow anguished,
> Fear and madness by my side,
> Suddenly, as though from Heaven,
> From the grave the stone was risen,
> And my soul was opened wide.[65]

And in the third of his *Hymns to the Night*, Novalis writes:

There came a twilight shutter, and all at once the bond of birth broke the Light's fetters. Fled was earthly splendor, and with it my grief. Condensed, sadness flowed away into a new and unfathomable world. Thou Night-inspiration, slumber of heaven, didst come over me: the region gently rose aloft and over the region hovered my released and newborn spirit. The mound became a cloud of dust and through the cloud I beheld the transfigured features of my Beloved. In her eyes reposed eternity...since then only have I felt everlasting, immutable faith in the heaven of the Night and in its Light, the Beloved.[66]

It was out of this profound initiatory experience that Novalis was lifted into a beholding of Sophie von Kühn as an aspect of Sophia, the Divine Feminine—leading to a single diary entry on June 29: "Christ and Sophia."[67] This realization found expression later in his *Hymns to the Night*: "The bosom of the world's love—Christ and Sophia,"[68] and was central throughout the creative outpouring that followed in the three and a half years until his death. Prophetically, Novalis writes in the second of the *Sacred Songs*: "Lo, there dawns the blessed hour," Christ is the one who sows "heavenly seeds."[69]

Novalis—Resurrection—Ascension of Jesus Christ

That Novalis had experienced a life-transforming meeting with the Christ is evident in the third of his *Sacred Songs*:

Thou need'st, like me, His Love full-flowing; The Comforter! For us He came. He died, with joy His Love bestowing On even those who caused His shame.	He brings new blood forever living Into the hard and dying bone; In answer to thy heart's free-giving He is eternally thine own.
He died, and now henceforth, forever Take thou His Love when sore oppressed, For thou canst trust Him, doubting never Within His arms to find a rest.	Whatever thou hast lost, He findeth, What thou hast loved He will restore; And dearer to thy heart He bindeth The gift His hand bestows once more.[70]

How might this meeting with the Christ be revealed as remembered by the stars? Is there a promise of *the cord of connection* in the birth horoscope of Novalis?

Referring to the comparison chart of Novalis's birth horoscope to the horoscope of the resurrection, might the alignment of the Moon (15♈08) in the central region of Aries at the birth of Novalis with

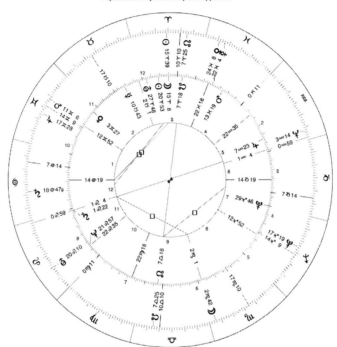

Comparison Chart

Outer - Geocentric
Resurrection of Jesus Christ
At Jerusalem, Latitude 31N46', Longitude 35E13'
Date: Sunday, 5/APR/33, Julian
Time: 5:35, Local Time
Sidereal Time 18:20:14, Vernal Point 2♈35'46"

Inner - Geocentric
Birth of Novalis
At Wiederstedt, Germany, Latitude 51N38', Longitude 11E30'
Date: Saturday, 2/MAY/1772, Gregorian
Time: 10: 8, Local Time
Sidereal Time 0:51: 0, Vernal Point 8♓26'12"

House System: Placidus, Zodiac: Sidereal SVP
Aspect set: Conjunction/Square/Opposition

the Sun's location (15♈39) at the time of the resurrection of Jesus Christ account for Novalis's connection to the Easter "constellation" of events in his own biography? As referred to already, the Moon in the birth horoscope reveals the conclusion of the weaving of the etheric body during the conception period, relating to the past soul karma of the individuality.

Moreover, in the horoscope of the day of the ascension (see appendix 2, chart 4), we see how the Moon returned forty days following the resurrection to *remember* the Sun's location at the resurrection, aligning once again with the location of the Moon at the time of Novalis's birth.

Might this connection to the Christ events be the deeper significance or meaning of the *pull of the cord of connection* to heaven indicated by the Sun's location in Pisces in the death horoscope of Novalis? Certainly the most central event of his life, the initiatory event experienced through the death of his beloved, Sophie von Kühn, gave birth to a spiritual baptism—this was the *pull of the cord of connection* drawing him ever nearer to the Divine. In the words of Novalis, "The old heaven must again come into view and with it the yearning for it, the living astronomy."[71]

5

RAPHAEL

FOLLOWING UPON THE THEME of baptism—with the purpose to reveal the primary focus of the Elijah individuality's mission from one life-time to the next—we shall follow the "golden thread" of connection, casting our gaze back from the incarnation as Novalis to the previous incarnation as the Renaissance painter Raphael and further back to the incarnation as John the Baptist. By examining the birth and death horoscopes of Raphael, we find an example of planetary alignment in relation to the event of the baptism in the Jordan.

Baptism in the Jordan—Raphael—Novalis

In the heliocentric/hermetic birth horoscope of Raphael, we find an otherwise hidden remembrance of the baptism in the Jordan.

In the comparison of the hermetic horoscopes between the birth and death of Raphael, the position of heliocentric *Mercury* at Raphael's *birth* is in sidereal Gemini (6Ⅱ24), very close to where heliocentric *Mercury* (7Ⅱ45) was in *conjunction* with *Saturn* (7Ⅱ18) at the baptism in the Jordan. This reveals that at Raphael's birth heliocentric Mercury had returned to almost exactly the same location as where it had been at the baptism in the Jordan (see comparison chart on page 46).

Raphael Birth h-Mercury (6Ⅱ24)—conjunct—
h-Mercury (7Ⅱ45)/h-Saturn (7Ⅱ18) Baptism

In Raphael's *death* horoscope, there is a heliocentric alignment of *Venus* with the heliocentric location of *Mercury* at his *birth*, on the

Comparison Chart

Outer - Tychonic
Baptism of Jesus Christ
At River Jordon, near Ono, Latitude 31N50', Longitude 35E33'
Date: Friday, 23/SEP/29, Julian
Time: 10: 0, Local Time
Sidereal Time 10: 0: 1, Vernal Point 2♈38'42"

Inner - Tychonic
Birth of Raphael
At Urbino/Italy, Latitude 43N44', Longitude 12E38'
Date: Friday, 28/MAR/1483, Julian
Time: 21:27, Local Time
Sidereal Time 10:25:25, Vernal Point 12♓27'56"

House System: Placidus, Zodiac: Sidereal SVP
Aspect set: Conjunction/Square/Opposition

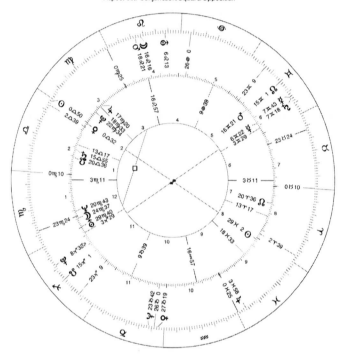

opposite side of the zodiac in sidereal Sagittarius (6♐48), thus recall-ing the *second "law" of reincarnation* concerning the interchange of the locations of heliocentric Venus/Mercury from death in one incar-nation to birth in the next incarnation—with the difference here that this alignment is in the *same incarnation.*

Heliocentric Venus' location in Sagittarius in the death horoscope of Raphael is then *remembered* in the next incarnation, appearing opposite in Gemini (7♊36) at the conception of Novalis—again echo-ing the location of the heliocentric Mercury/Saturn conjunction at the baptism in the Jordan.

In the following summary of these alignments all the planetary positions are heliocentric and sidereal:

Comparison Chart

Outer - Tychonic	Inner - Tychonic
Death of Raphael	Birth of Raphael
At Rome, Italy, Latitude 41N53', Longitude 12E30'	At Urbino/Italy, Latitude 43N44', Longitude 12E38'
Date: Friday, 6/APR/1520, Julian	Date: Friday, 28/MAR/1483, Julian
Time: 12: 0, Local Time	Time: 21:27, Local Time
Sidereal Time 1:36:42, Vernal Point 11 Ⅹ56'59"	Sidereal Time 10:25:25, Vernal Point 12 Ⅹ27'56"

House System: Placidus, Zodiac: Sidereal SVP
Aspect set: Conjunction/Square/Opposition

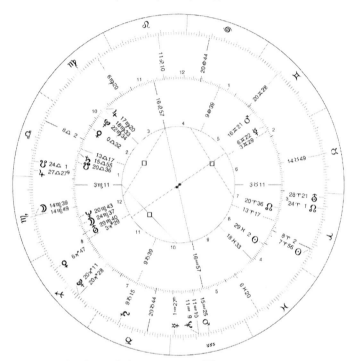

Raphael Birth ♄-Mercury (6Ⅱ24)—opposite—
Raphael Death ♄-Venus (6♐48)

Baptism in the Jordan ♄-Mercury (7Ⅱ45)—conjunct—
Raphael Birth ♄-Mercury (6Ⅱ24)
Raphael Birth ♄-Mercury (6Ⅱ24)—opposite—
Raphael Death ♄-Venus (6♐48)
Raphael Death ♄-Venus (6♐48)—opposite—
Novalis Epoch ♄-Venus (7Ⅱ36)

Raphael and the Italian Renaissance:

Raphael entered earthly life at the time of the full flowering of
the Italian Renaissance. The significance of this thread was visible
to the contemporaries of Novalis, who himself is considered to have

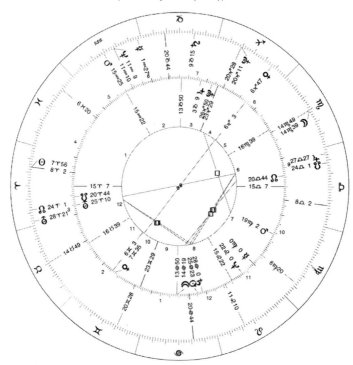

Comparison Chart

Outer - Tychonic
Death of Raphael
At Rome, Italy, Latitude 41N53', Longitude 12E30'
Date: Friday, 6/APR/1520, Julian
Time: 12: 0, Local Time
Sidereal Time 1:36:42, Vernal Point 11 ⅩƆ56'59"

Inner - Tychonic
Epoch of Novalis
At Wiederstedt, Latitude 51N38', Longitude 11E30'
Date: Friday, 9/AUG/1771, Gregorian
Time: 21: 7, Time Zone GMT
Sidereal Time 19: 5:17, Vernal Point 8 Ⅹ26'49"

House System: Placidus, Zodiac: Sidereal SVP
Aspect set: Conjunction/Square/Opposition

Raphael Death ♄-Venus (6♐48)—opposite—
Novalis Epoch ♄-Venus (7Ⅱ36)

been important to the seeding and rise of the Romantic period in European literature. On July 13, 1799, two years after the death of Sophie von Kühn, Novalis made the acquaintance of Ludwig Tieck, whose importance was significant in that it was through Tieck that Novalis came to know the work of the German mystic Jacob Boehme, whose writings revealed the *reality* of Sophia, and thus further nourished Novalis's thirst for the Divine Feminine: "*Among all the mystics none placed the mystery of the Virgin Sophia so radiantly at the heart of Novalis's thinking as did Boehme.*"[72]

Might there be seen a *seed* of connection to the Divine Feminine following from Novalis's prior life as Raphael? Among Novalis's circle of friends, Raphael was considered the prototype of the Renaissance, the representative of the flowering of true culture. We find in the words of Novalis's friend Tieck a glimmer of recognition:

> With the intuitive vision of true friendship, Tieck divined the spiritual background of themes in Novalis's creative outpouring of literature. Unconsciously at first, and then, more and more consciously, Tieck felt as though something of Raphael's Florence and Rome had reawakened, had, in the early Romantic Movement, become transformed from image to word.[73]

In the final pages of his diary, shortly before his death from tuberculosis on March 25, 1801, Novalis wrote: *"Have I not from time immemorial chosen all my own destinies?"*[74]

Into what world did Raphael choose to incarnate? In what garden did he choose to flower? Raphael was born in Urbino, Italy, on the 28th of March 1483. According to historical accounts, to be born in Urbino was a privilege, as Urbino had become an outstanding cultural center by the time of Raphael's birth and therefore was open to all the ideas and speculations of the Renaissance. It is clear from these accounts that Urbino was considered one of the most civilized places in Europe. Thus Raphael was born into these most cultivated of circles, circles in which the standard for behavior took on a *moral* context. There was a code of behavior for the gentleman, who was to be a man of honor and decency:

> ... who is more than proficient in letters, at least in those studies that we call the humanities, and is acquainted not only with Latin, but with Greek as well, on account of the many and various things that in that tongue are sublimely written. Let him be versed in the poets and no less in the orators and historians, and skilled too in the writing of verse and prose, more especially in this our own language." [75]

The ideals of Urbino stayed with Raphael, providing a background for both his art and life.

Raphael's Birth Date: Unveiling the Mystery

Various biographies of Raphael indicate that he was born and died on the same day, and since his death date is known to be Good Friday, April 6, 1520 and that he was 37 years old at his death, sometimes his birth date is given as April 6, 1483. However, in terms of the Christian year, the fact that he was born and died on the same day points to his birth having been on Good Friday in the year 1483, which equates with March 28, 1483. This signifies that at his death he was 37 years and 9 days old, which agrees with the statement that "he died shortly after his 37th birthday."[76] Moreover, from the standpoint of astrosophical research, March 28, 1483, is confirmed as the date of birth.

Now there remains the question as to his time of birth. According to Vasari's book, *Lives of the Artists*, published in Florence in 1550, just thirty years after Raphael's death: [Raphael was born on] Good Friday of the year 1483 at 3 o'clock at night. Here it has to be born in mind that in Italy at the time of the Renaissance night began at sunset, or rather at dusk, signifying that Raphael was born at about 9:30 PM—three hours after dusk—on March 28, 1483, in Urbino, Italy.

Raphael: Determining the Epoch

Given the time of birth, we can apply the hermetic rule to determine Raphael's epoch/conception. One possibility for the epoch is July 10, 1482 at 1:23 AM local time, which equates with 0:33 AM Greenwich Mean Time (GMT). Let us consider a comparison of the *hermetic epoch* with the *hermetic death* chart of Raphael (opposite). Here again we are reminded of the second *"law"* of reincarnation, with *Venus* in sidereal Leo (0♌22) at the conception of Raphael in *alignment* on the opposite side of the zodiac with *Mercury* in sidereal Aquarius (1♒29) at Raphael's death.

It should be noted that, as we do not know the time of Raphael's death, the death horoscope is computed for midday on his death date, signifying that heliocentric *Mercury* in sidereal Aquarius (1♒29) is an "average position" on that day. If Raphael died early in the morning, not long after midnight, heliocentric Mercury would have

Comparison Chart

Outer - Tychonic
Death of Raphael
At Rome, Italy, Latitude 41N53', Longitude 12E30'
Date: Friday, 6/APR/1520, Julian
Time: 12: 0, Local Time
Sidereal Time 1:36:42, Vernal Point 11 ♓56'59"

Inner - Tychonic
Epoch of Raphael
At Urbino/Italy, Latitude 43N44', Longitude 12E38'
Date: Wednesday, 10/JUL/1482, Julian
Time: 0:33, Time Zone GMT
Sidereal Time 21:10: 6, Vernal Point 12 ♓28'31"

House System: Placidus, Zodiac: Sidereal SVP
Aspect set: Conjunction/Square/Opposition

been located at 29♑36 in sidereal Capricorn. However, if he died late at night, shortly before midnight, heliocentric Mercury would have been located in Aquarius at 3♒22. In this way we can estimate that heliocentric Mercury was somewhere in the four degree range between the end of Capricorn (29♑36) and the beginning of Aquarius (3♒22) at the time of Raphael's death, with 1♒29 being the noon average between these two locations.

> *Heliocentric Mercury/Venus interchange in the hermetic comparison horoscope: Raphael Epoch ♄-Venus (0♌22)— opposite—Raphael Death ♄-Mercury (1♒29)*

The most striking thing about the comparison of the *hermetic epoch* and *death* charts is the *interchange* of the *Sun* and *Saturn*.

For example, the Sun was in sidereal Aries (7♈57) at Raphael's death exactly *opposite* heliocentric Saturn in sidereal Libra (7♎43) at his conception. Correspondingly *heliocentric* Saturn in sidereal Capricorn (9♑15) in Raphael's death horoscope was *aligned* with the *Sun* on the *opposite* side of the zodiac in Cancer (8♋21) at Raphael's conception.

Heliocentric Saturn/Sun interchange in the hermetic comparison horoscope:

Raphael Epoch Sun (8♋21)—opposite—
  Raphael Death ♄-Saturn (9♑15)
Raphael Epoch ♄-Saturn (7♎43)—opposite—
  Raphael Death Sun (7♈57)

This is in contrast to the death horoscope of Novalis wherein *geocentric* Saturn (25♋41) was in *conjunction* with the Epoch Sun (25♋24) in sidereal Cancer at Novalis's conception—in exact alignment with the Sun's location (25♋24) in the birth horoscope of Mary of Nazareth.[77]

Novalis Epoch Sun (25♋24)—conjunct—
  Novalis Death g-Saturn (25♋41)
Novalis Epoch Sun (25♋24)—conjunct—
  Mary of Nazareth Birth Sun (25♋24)

Raphael's Legacy

What do we know of the death of Raphael? What was the legacy of this life as Raphael and how did it serve Novalis in the next incarnation? When we contemplate Raphael's life, it can be seen that this individuality held true to the words of Christ, *"Elijah does come, and he will restore all things"* (Matthew 17:11). We know that Raphael went to Rome in 1508 and was commissioned by Pope Julius II, who had begun the work of renovating the Vatican Palace and rebuilding St. Peter's, intending to renew and reconfirm the grandeur of Rome. Julius noted immediately how Raphael "set to work with such serene confidence, painting in so delicate and sweet a manner, that Pope Julius caused all the pictures of the other masters, both ancient and modern, to be destroyed, that Raphael might have the sole credit for all the works here performed down to the present time."[78] Thus with papal consent Raphael painted a series of "picture-poems" in the Vatican Stanze. For example, the paintings in the Stanza della Segnatura were intended to demonstrate the cultural and ideological legacy of the Neoplatonic ideal, the central theme of this series of paintings being the True, the Good, and the Beautiful. Here he achieved a sublime expression of form creating an impression of the "divine imprint" of the harmony of the universe.

As a legacy to humanity, holding true to the restoration of truth, the *Disputa* (the dispute over the Holy Sacrament, the famous fresco representing the "Triumph of the Holy Eucharist") and the fresco depicting the *School of Athens* record the truth coming to expression in the theological and philosophical realms of the developing consciousness of humanity.

Later, under Pope Leo X, Raphael became the chief architect of St. Peter's and changed the central plan of St. Peter's into a cruciform, redeeming its form for posterity.

With his final painting, The *Transfiguration*, Raphael left humanity with a prophetic vision of the transfigured Christ *appearing* between Moses and Elijah as though coming again *out of the clouds* (see front cover). In this magnificent work, Raphael is recording an historical event from the time of Christ in a manner that is entirely prophetic, in that it depicts Christ in the form in which he is to return through his second coming. As we shall consider in more detail in the last chapter, it is through his painting of the *transfiguration* that Raphael emerges as a true herald for the greatest event of our time. [79]

6

DEATH RAPHAEL – BIRTH NOVALIS

CONSIDERING THE CONCORDANCE BETWEEN the epoch/conception and death charts of both Raphael and Novalis, we can be fairly certain that the respective epoch dates that we have been exploring are correct. Having found (with a high degree of certainty) these epoch/ conception dates, we are now in a position to do chart comparisons between Raphael and Novalis for all three horoscopes (epoch, birth, death) both geocentric and heliocentric/hermetic, which means a total of eighteen chart comparisons (9 geocentric and 9 hermetic). With so many chart comparisons, obviously we are looking for only the most striking connections in our search for the true and deeper foundations of astrology. So far, horoscopes have been included in the text; from now on, the reader should consult appendix 2 for all details referring to chart computations of horoscopes.

The First "Law" of Astrological Reincarnation

Let us begin with the comparison of the *death* horoscope of Raphael with the *birth* horoscope of Novalis, starting with the *first* "law" of astrological reincarnation regarding the *angular relationship between the Sun and Saturn* between the *death* in one incarnation and *birth* in the following incarnation (see comparison chart 10 in appendix 2). We discover that the angle between the Sun and Saturn at the death of Raphael was 82°59', whereas the angle between the Sun and Saturn at the birth of Novalis was 100°27'. Obviously, these two angles are not the same, yet their sum is 183°26', which is close to 180 degrees. Because the two angles are (approximately) complementary (with respect to 180 degrees), the first "law" is fulfilled.[80]

Hermetic comparison:

For the sake of comparison, let us consider the corresponding angles in the *hermetic* charts (see appendix 2, chart 11):

> *The angle between the Sun and heliocentric-Saturn*
> *at death Raphael = 88°42'*
> *The angle between the Sun and heliocentric-Saturn*
> *at birth Novalis = 106°39'*

Again, these two angles are not the same. Their sum is 195°21', so they cannot be regarded as complementary with respect to 180 degrees, since the difference (15°21') is too large.

Therefore in the case of the reincarnation of Raphael as Novalis, it is solely in the comparison of the geocentric charts that the first "law" of astrological reincarnation is fulfilled.

Uranus Alignment (Geocentric and Hermetic)

In terms of planetary alignments, the most striking is the *geocentric* alignment of *Uranus* at Novalis's *birth* (27♈49) with Uranus at Raphael's *death* (27♈22). The *heliocentric* (hermetic) alignment of *Uranus* is even more exact: with Uranus (28♈21) at Raphael's *death* and Uranus (28♈10) at Novalis's *birth* in conjunction, only eleven minutes apart. The Uranus alignment occurs in other reincarnation examples as well.[81]

Summarizing this discovery from the geocentric and hermetic comparison charts in appendix 2, where g = geocentric and h = heliocentric/hermetic:

> *Raphael Death g-Uranus (27♈22)—three Uranus orbits—*
> *Novalis Birth g-Uranus (27♈49)*
> *Raphael Death h-Uranus (28♈21)—three Uranus orbits—*
> *Novalis Birth h-Uranus (28♈10)*

What does this *Uranus alignment* signify? It could be hypothesized that the reincarnating soul chooses to reincarnate upon the completion of a certain number of orbits of Uranus around the sidereal zodiac, when Uranus returns again to the same place in the zodiac where it had been at death in the preceding incarnation. Since 252 years elapsed between Raphael's death in 1520 and the birth of Novalis in 1772, and since Uranus makes one orbit of the zodiac in 84 years, this means that Novalis waited to reincarnate upon the completion of three (3 x 84 = 252) orbits of Uranus around the zodiac.

The Second "Law" of Astrological Reincarnation

Looking now at the *hermetic* chart comparisons of the *death* of Raphael with the *birth* of Novalis (see comparison chart 11 in appendix 2), let us see if *the second "law" of astrological reincarnation* is fulfilled.

We can observe that there is a "loose" *opposition* between Mercury in Aquarius at Raphael's death and Venus in Leo at Novalis's birth:

> *Death Raphael h-Mercury (1♒29)—opposite—*
> *Novalis Birth h-Venus (15♌28)*[82]

However, we find a fairly exact *alignment* of *heliocentric Mars* in the *death* horoscope of Raphael [*Mars* in Aquarius at 15♒25] and in the *birth* horoscope of Novalis with *heliocentric Mars* at 16♒54. In this example the alignment of *heliocentric Mars* could be seen as a kind of *substitute* for the *heliocentric Mercury* and/or *heliocentric Venus alignment of the second "law."*

Alignment of heliocentric Mars:

> *Raphael Death h-Mars (15♒25)—conjunct—*
> *Novalis Birth h-Mars (16♒54)*

The Occurrence of Interchange

The interchange of the *Sun and the Moon's Node* (Moon's nodal axis) is striking in the comparison chart:

Raphael Death Sun in sidereal Aries at 7♈57 aligns with *Novalis Birth Moon's Descending Node* at 7♈18 (see comparison chart 10 in appendix 2).

Correspondingly, the *Moon's Ascending Node* at Raphael's *death* was at 24♈01 compared to the location of the *Sun* at 20♈53 at the *birth* of Novalis. (*This interchange is the same in both the geocentric and the hermetic charts.*)

On the basis of decades of research I can say that this kind of interchange is characteristic of authentic reincarnation examples.

Sun/Moon's Node Interchange: Geocentric and Hermetic:

Raphael Death Sun (7♈57)—conjunct—Novalis Birth Descending Moon Node (7♈18)
Raphael Death Ascending Moon Node (24♈01)—conjunct— Novalis Birth Sun (20♈53)

Mars: Geocentric Alignment

In the *geocentric* chart comparison between the *death* of Raphael and the *birth* of Novalis, the geocentric positions of *Mars* are in the same sign and are not too far apart. The proximity means that this can be considered a *loose alignment*: Raphael Death Mars in sidereal Pisces at (7♓14) and Novalis Birth Mars again in sidereal Pisces at (13♓20).

Raphael Death g-Mars (7♓14)—conjunct— Novalis Birth g-Mars (13♓20)

Here it is evident, though, that the more exact Mars alignment is to be found in the hermetic chart comparison discussed earlier:

Alignment of heliocentric Mars:

Raphael Death h-Mars (15♒25)—conjunct— Novalis Birth h-Mars (16♒54)

Pluto: Geocentric Alignment

It is noteworthy that Pluto had just completed one orbit (orbital period ca. 247 years) between the death of Raphael and the birth of

Novalis (252 years). At the death of Raphael geocentric Pluto was located in sidereal Sagittarius at 21♐52 compared to the birth of Novalis—with geocentric Pluto again in sidereal Sagittarius (29♐46).

> *Raphael Death g-Pluto (21♐52)—one orbit of Pluto—*
> *Novalis Birth g-Pluto (29♐46)*

The difference between these two zodiacal locations of Pluto is eight degrees, which is the outermost limit of what could be described as a "loose conjunction"—the normal range or orb for a conjunction being three degrees.

Ascendant-Descendant Axis: Geocentric Alignment with Saturn

Most striking, however, is the alignment of the *Ascendant-Descendant axis* at Novalis's *birth* with geocentric *Saturn* at Raphael's *death*: Raphael Death Saturn (14♑58) compared to the *birth* of Novalis: Ascendant (14♋19), signifying that Novalis's *Descendant* in sidereal Capricorn *pointed toward* geocentric *Saturn* at death in the preceding incarnation. *This is interesting in view of the traditional association of the Descendant with death.* In this example we gain a glimpse of the concrete significance of this traditional association. So it is that astrological reincarnation research is able to uncover and reveal the deeper background to astrological teaching. It should be added that the alignment of the *Ascendant–Descendant axis* with *Saturn* from *death* in the *preceding incarnation* is something that occurs not infrequently, so that *it almost warrants being called a third "law."*

Third "law": Ascendant-Descendant axis alignment with geocentric Saturn:

> *Raphael Death g-Saturn (14♑58)—opposite—*
> *Novalis Birth Ascendant (14♋19)*
> *Raphael Death g-Saturn (14♑58)—conjunct—*
> *Novalis Birth Descendant (14♑19)*
> *Raphael Death g-Saturn (14♑58)—opposite—*
> *Novalis Epoch Moon (14♋19)*

The third and last of the above alignment equations needs some clar-ification. According to the *hermetic rule the Ascendant–Descendant axis at birth aligns with the Moon at epoch*. This signifies that the *Moon* at epoch/conception has a tendency to *align* with *Saturn* at *death* in the preceding incarnation. In this case at the death of Raphael geo-centric Saturn was located in sidereal Capricorn (14♑58) in *opposi-tion* to Novalis's conception Moon in sidereal Cancer (14♋19). What other alignments do we find between the death horoscope of Raphael and the conception horoscope of Novalis?

Baptism in the Jordan

As described earlier, we are reminded of *the second "law" concern-ing Venus and Mercury* revealed in the *hermetic* charts of Raphael's *death* and Novalis's *epoch/conception*:

> *Raphael Death ♄-Venus (6♐48)—opposite—*
> *Novalis Epoch ♄-Venus (7♊36)*
> *Raphael Birth ♄-Mercury (6♊24)—conjunct—*
> *Novalis Epoch ♄-Venus (7♊36)*
> *Baptism in the Jordan: ♄-Saturn (7♊18) conjunct*
> *♄-Mercury (7♊45)*

In the case of the above line relating to the baptism in the Jordan, this is simply a conjunction, heliocentrically, between the planets Saturn and Mercury, and is not an alignment, which is why it is written "conjunct" rather than "— conjunct —," which indicates an alignment.

Here we have a series of *alignments* all linking back to the *helio-centric conjunction of Mercury and Saturn at the baptism in the Jordan*, when heliocentric Saturn (7♊18) was conjunct heliocentric Mercury (7♊45) in sidereal Gemini.

In the case of the individuality of John the Baptist, reincarnating as Raphael and later as Novalis, we see heliocentric Mercury and/or heliocentric Venus continually aligning with this same position—at epoch, birth, death—in memory of the baptism in the Jordan. This is the deeper background to the heliocentric Mercury/Venus align-ments in the case of this individuality. This is just one example of a "nugget of gold" we have uncovered through astrological reincarnation research, offering a glimpse into profound karmic (and cosmic) mysteries.

Unveiling the Mystery:

Before proceeding with further chart comparisons, let us summarize what we have observed so far. First, all three charts—epoch, birth, and death—are highly significant in terms of planetary alignments in the sidereal zodiac.

Second, the second "law" of astrological reincarnation, when it is fulfilled, seems to point to something of overriding significance in the karmic biography of the individuality, e.g. in the case of John the Baptist–Raphael–Novalis, it is the event of the baptism in the Jordan that is of supreme significance.

Third, the slow-moving outer planets—Uranus, Neptune, Pluto—are evidently significant in that a complete number of orbits often take place between incarnations, e.g. between the birth of John the Baptist and Raphael six complete orbits of Pluto, and between the death of Raphael and the birth of Novalis three complete orbits of Uranus.

7

CHART COMPARISONS: RAPHAEL–NOVALIS

BIRTH RAPHAEL–BIRTH NOVALIS: MOON'S NODAL AXIS

S O FAR WE HAVE compared the death of Raphael with the birth of Novalis. Let us now compare the birth horoscopes of Raphael with the birth horoscopes of Novalis (the plural "horoscopes" signifies the geocentric and the heliocentric/hermetic horoscopes of birth). Attention has already been drawn to the loose alignment between the *Sun* at the birth of Novalis with the *Moon's Ascending Node* at the death of Raphael. However, this alignment is *exact* between the birth of Raphael and the birth of Novalis (see appendix 2, chart 9):

> *Raphael Birth Ascending Moon Node in*
> *   sidereal Aries (20♈36)*
> *Novalis Birth Sun in sidereal Aries (20♈53)*

Keeping in mind the corresponding Sun/Moon Node alignments discussed earlier:

> *Novalis Epoch Descending Moon Node (20♈44):*
> *Novalis Birth Descending Moon Node (7♈18)*
> *Raphael Death Sun in sidereal Aries (7♈57)*

Here our attention is drawn to something special in the tendency of the *Moon's nodal axis* to align with the *Sun or with itself*, when we recall the alignments between the death of Raphael and the birth of Novalis:

> *Raphael Death Ascending Moon Node (24♈01)*
> *Novalis Birth Sun in sidereal Aries (20♈53)*

Biographical Homily: Raphael—Novalis:

The Moon's nodal axis requires 18.61 years to return to the same location as at the time of one's birth. Because the Moon's Node marks the zodiacal location of the Moon's crossing the line of the ecliptic, this rhythm has been observed to mark times of great import to the individuality, offering an opening or access to the more expanded perspective of the realm of the fixed stars. It is interesting to note that Raphael's death, following shortly after his thirty-seventh birthday, falls within this rhythm, since two Moon Node intervals of 18.61 years amount to thirty-seven years and two months. The death of Raphael at the age of thirty-seven clearly relates to the Moon Node rhythm, although his age at death is an approximation rather than an exact fulfillment of two Moon Node cycles.

In terms of the *ripening* of the soul for the holy encounter with death (as a birth into spiritual realms[83]), Raphael was working on his painting of the *Transfiguration* just before his death. As he himself was the reincarnated Elijah, Raphael was mystically present or attuned to the actual event of the Transfiguration. At the time of Christ's transfiguration, the Moon's nodal axis was aligned with the two royal stars, Aldebaran (15° Taurus) and Antares (15° Scorpio). (See David Tresemer's *Star Wisdom and Rudolf Steiner* concerning Aldebaran and Antares.) A further impression of the magnitude of the radiance of the actual transfiguration is indicated by the position of the Sun at 14° Aries, which was in the region of its *exaltation* (19° Aries). At the Transfiguration the *exalted* Son of Man appeared in radiant light between Moses and *Elijah*.

This lunar node rhythm appears again in the life of the young Novalis who, in the conventional way of thinking, was thought to be rather slow and unable to master learning up until his ninth year, when following a mysterious bout of dysentery he emerged with a remarkable mental awakening, indicating the availability of more of his Sun-like "I" nature to shine through. The ninth year marked the halfway point of the first lunar node cycle in his biography.

Moon / Saturn: Geocentric Alignment

We may also note the geocentric alignment of the Moon at the birth of Novalis with Saturn at the birth of Raphael—on *opposite* sides of the zodiac (see comparison chart 9 in appendix 2):

> *Raphael Birth g-Saturn (17♎48)—opposite—*
> *Novalis Birth Moon (15♈08)*

It is a matter of looking for tendencies. Thus we may recall the *alignment* of the Moon at Novalis's epoch with geocentric Saturn at Raphael's death:

> *Novalis Epoch Moon (14♋19)—opposite—*
> *Raphael Death g-Saturn (14♑58)*

Obviously the *relationship* of the *Moon* to *Saturn* is something important to our investigation.

Birth Raphael—Conception Novalis

Turning now to compare Raphael's birth horoscope with that of Novalis's conception, again the most striking alignment is that of the *Moon's nodal axis* (see appendix 2, comparison chart 6):

> *Raphael Birth Ascending Moon Node (20♈36)—conjunct—*
> *Novalis Epoch Descending Moon Node (20♈44)*

Now comparing the *hermetic* charts, the *Descendant* at Novalis's epoch/conception points to heliocentric Saturn at Raphael's birth (see the hermetic comparison chart 7 in appendix 2):

> *Raphael Birth h-Saturn (15♎55)—conjunct—*
> *Novalis Epoch Descendant (15♎07)*

The heliocentric alignment of *Mercury* and *Venus* has already been referred to:

> *Raphael Birth h-Mercury (6♊24)—conjunct—*
> *Novalis Conception h-Venus (7♊36)*

Extending this heliocentric alignment back in time, we return to the alignment with *h-Mercury* at the Baptism in the Jordan (7♊45).

Death Raphael—Conception Novalis

As discussed earlier, a *Venus/Venus* alignment occurred, evident when we compare the *hermetic* chart of Raphael's death with that of Novalis's epoch/conception:

> *Raphael Death h-Venus (6♐48)—opposite—*
> *Novalis Epoch h-Venus (7♊36)*

Now looking at the *geocentric* comparison chart of Raphael's epoch/conception with Novalis's death, we find geocentrically an interchange of the *Moon* and *Venus:*

> *Raphael Epoch Moon (3♉11)—loosely conjunct—*
> *Novalis Death g-Venus (28♈10)*
> *Raphael Epoch g-Venus (17♋26)—conjunct—*
> *Novalis Death Moon (16♋25)*

In the corresponding *hermetic* chart comparison, we find again a *Venus/Venus* alignment:

> *Raphael Epoch h-Venus (0♌22)—conjunct—*
> *Novalis Death h-Venus (28♌44)*

Here also the location of h-Saturn should be noted: *Novalis Death h-Saturn (0♌13).*

There are two other noteworthy alignments in this *hermetic* chart comparison, which are striking, due to their exactness:

> *Raphael Epoch h-Mars (16♑28)—opposite—*
> *Novalis Death Moon (16♋25)*
> *Raphael Epoch h-Jupiter (27♌29)—conjunct—*
> *Novalis Death h-Mercury (27♌28)*

At this juncture, the collection of research material has grown considerably. Before continuing, let us remind ourselves that our goal is to stand upon a firm foundation in the new science of astrology based on astrological reincarnation research, and that *empirical* evidence is important for this. It follows that the gathering of research material is necessary for the discovery of astrological *reincarnation "laws" and principles*, which surely will begin to emerge from our contemplation of the evidence amassed. Bearing this in mind, let us continue with our study of the reincarnation example, Raphael–Novalis.

Raphael Conception—Novalis Birth

Now we come to a comparison of the epoch/conception horoscope of Raphael with the birth horoscope of Novalis. In the *geocentric* chart comparison there are two alignments involving *Jupiter*:

> *Raphael Epoch g-Jupiter (20♌18)—conjunct—*
> *Novalis Birth g-Neptune (21♌57)*
> *Raphael Epoch g-Mars (6♒55)—conjunct—*
> *Novalis Birth g-Jupiter (7♒24)*

In the *hermetic* chart comparison there is the following alignment:

> *Raphael Epoch h-Saturn (7♎43)—conjunct—*
> *Novalis Birth Ascending Moon Node (7♎18)*

Raphael—Novalis Conception Charts

In the comparison of the epoch/conception charts of Raphael and Novalis, first let us consider the *geocentric* level:

> *Raphael Epoch g-Mercury (3♋53)—conjunct—*
> *Novalis Epoch g-Venus (5♋48)*
> *Raphael Epoch Sun (8♋20)—conjunct—*
> *Novalis Epoch g-Venus (5♋48)*
> *Raphael Epoch g-Venus (17♋ 6)—conjunct—*
> *Novalis Epoch Moon (14♋19)*

Now let us consider the *heliocentric/hermetic* level:

> *Raphael Epoch h-Mars (16♑28)—opposite—*
> *Novalis Epoch Moon (14♋19)*
> *Raphael Epoch h-Jupiter (27♌29)—conjunct—*
> *Novalis Epoch h-Mercury (0♍02)*

Raphael Birth—Novalis Death

There still remains the comparison between Raphael's birth and Novalis's death horoscopes. The geocentric comparison reveals the following alignments:

> Raphael Birth g-Jupiter (14♍44)—opposite—
>   Novalis Death Sun (12♓24)
> Raphael Birth Moon (24♏37)—opposite—
>   Novalis Death g-Mars (26♉41)

Raphael—Novalis: Death Chart Comparison

Finally, comparing the death chart of Raphael with the death horoscope of Novalis, on the *geocentric* level we find:

> Raphael Death g-Saturn (14♑58)—opposite—
>   Novalis Death Moon (16♋25)
> Raphael Death g-Mercury (20♓36)—conjunct—
>   Novalis Death g-Mercury (21♓03)

The exact alignment of *geocentric Mercury* here is striking (the orb is less than ½ degree). However, without having a meaningful number of other reincarnation examples to compare this and the other alignments (listed above and below), it is not possible at this stage to draw definite conclusions.

What is being outlined here in this book is a *methodology* as an example for other researchers. Hopefully, in the course of time the empirical findings of astrological reincarnation research will be sufficient enough to discern definite patterns of alignments from one incarnation to the next, or even between incarnations widely separated from one another in time with one or more incarnations in between.

This book is intended as a work outlining possible research procedures for the future of astrological reincarnation research, taking as our point of departure the *first* concrete reincarnation example given publicly by Rudolf Steiner, whose real task was indicated by Walter Johannes Stein according to his conversation with Rudolf Steiner in The Hague, Holland, in 1922: "Destiny working over from one earthly life to another stood before his soul. To work this out in concrete knowledge was *his* mission."

However, as Rudolf Steiner himself said: "I made many attempts to carry out my own destiny, but because of outside resistance I could not do it."[84] What Rudolf Steiner meant is that, as he experienced it, there were very powerful forces at work to hinder knowledge of reincarnation and karma coming into the world at that time in the first part of the twentieth century. Of course, these forces are still at work now in a somewhat different way in the twenty-first century. Notwithstanding the activity of such forces in the contemporary world, this book is intended as a contribution to the pioneering work of Rudolf Steiner in the realm of reincarnation research.

Returning now to the geocentric comparison chart of the death of Raphael with that of Novalis, let us note that the *geocentric opposition* between *Venus* and *Neptune* at the death of Novalis aligns with *geocentric Uranus* at the death of Raphael, and also that *geocentric Pluto* at the death of Novalis aligns with *geocentric Neptune* at the death of Raphael.

Here we see the alignments of the outer planets Uranus, Neptune, and Pluto playing a role in the geocentric chart comparison between death in one incarnation and death in the next incarnation:

> *Raphael Death g-Uranus (27♈22)—conjunct—*
> *Novalis Death g-Venus (28♈10)*
> *Raphael Death g-Uranus (27♈22)—opposite—*
> *Novalis Death g-Neptune (27♎08)*
> *Raphael Death g-Neptune (12♒43)—conjunct—*
> *Novalis Death g-Pluto (12♒52)*

Raphael Death—Novalis Death (Hermetic)

In the *hermetic* chart comparison of Raphael's death chart and Novalis's death horoscope we find a fulfillment of the *second "law"*:

> *Raphael Death h-Mercury (1♒29)—opposite—*
> *Novalis Death h-Venus (28♌44)*

Thus far we have carried out all possible (nine) chart comparisons between Raphael and Novalis—two successive incarnations of the same individuality—and arrived at the above alignments.[85] However, we have not yet paid attention to *aspect metamorphoses,* defined as the recurrence of planetary aspects from one incarnation to the next.[86]

Aspect Metamorphosis

An aspect metamorphosis is defined as an aspect between two (or more) planets in one incarnation that recurs—either as the same aspect or as a metamorphosis of this aspect—in the next incarnation. For example, a conjunction between two planets at birth in one incarnation could become an opposition between these two planets at conception in the next incarnation—bearing in mind the fundamental law of astrological reincarnation indicated in figure 2 at the start of chapter 3, where there is a relationship between birth in the one incarnation and conception in the next incarnation. For example, it is noteworthy that there was a geocentric conjunction of Jupiter and Pluto at Raphael's birth and again a geocentric conjunction of Jupiter and Pluto at the epoch/conception of Novalis.

Aspect Metamorphosis Example: (Geocentric)
Raphael Birth: Jupiter/Pluto conjunction—Novalis Epoch:
Jupiter/Pluto conjunction

However, whereas the orb of the Jupiter/Pluto conjunction at the conception of Novalis is less than two degrees, the Jupiter/Pluto conjunction at Raphael's birth is only a *loose conjunction*, since the orb is almost eight degrees. It is extraordinary that there are no other notable aspect metamorphoses apparent in all the nine comparison charts between Raphael and Novalis—actually eighteen comparison charts, taking account of the geocentric and heliocentric/hermetic levels. This fact would seem to indicate that aspect metamorphoses play only a minor role (if any) in astrological reincarnation, and that the major factor at work in the soul's choice of horoscope when reincarnating is the principle of planetary alignments in the sidereal zodiac (both geocentric and heliocentric/hermetic). This is a cautious preliminary conclusion based on the analysis of comparison charts between two successive incarnations of the Elijah individuality as Raphael and Novalis. Obviously, more research—using many other reincarnation examples—needs to be carried out before coming to a firm conclusion regarding the significance (or lack of significance) of aspect metamorphoses.

In the next chapter let us consider the discovery of the *metamorphosis* from the incarnation of Elijah as John the Baptist to the life and times of St. Elisabeth of Hungary, Raphael's prior incarnation as a "saintly nun."

8

St. Elisabeth – Raphael – Novalis

IN ADDITION TO THE incarnations Raphael and Novalis, there are frequent references by Rudolf Steiner to the two earlier incarnations: Elijah and John the Baptist. For astrological reincarnation research purposes, the incarnation as Elijah lies beyond our sphere of research, since we do not know anything about the dates of birth and death of Elijah, except that he lived in the ninth century B.C.E. The same applies to a still earlier incarnation of Elijah, indicated by Rudolf Steiner, as Phinehas, who lived at the time of Moses in the thirteenth century B.C.E. (Numbers 25:6–12). Are there any further incarnations of this individuality indicated by Rudolf Steiner?

There is an incarnation that he indicated indirectly in the following quotation by Marie Steiner (Marie von Sivers):

> It was the time when he [Rudolf Steiner] encouraged me more and more to do recitals. I tried to arrive at the essence of Novalis. I told him that it was not easy for me, and that I had not yet found the key to Novalis's being. He advised me to steep myself in the mood of the saintly nuns.... One day enlightenment came. The figure of Raphael stood before me.... I said to Rudolf Steiner, "The nuns did not help, but another did: Raphael. Now Novalis has become quite transparent for me." A radiance appeared on Rudolf Steiner's tender face. Some days later he revealed to us for the first time the mystery of Novalis–Raphael–John the Baptist–Elijah.[87]

Clearly, Rudolf Steiner had in mind an earlier incarnation of Novalis as a "saintly nun." You can imagine my joy when I read the

above quotation in the 1980s, having already found in the 1970s
the preceding incarnation of Raphael–Novalis as St. Elisabeth of
Hungary (also known as St. Elisabeth of Thüringia). Before going on
to consider this in the light of astrological reincarnation research, let
us contemplate the life of this extraordinary woman.

St. Elisabeth of Thüringia

The year 2007, the year in which the first draft (from 2003) of this
book was revised, marked the eight-hundredth anniversary of the birth
of St. Elisabeth. She was born in early July of the year 1207, and we shall
return to consider the exact date later. Her father was King Andreas II
of Hungary, and her mother, Gertrude, was renowned for her piety
and strength of purpose. Elisabeth was born at Sárospatak in Hungary.
When she was four years old, Landgrave Hermann I of Thüringia sent
messengers to Hungary to ask if Elisabeth would become the future
bride of his eldest son, Hermann. King Andreas and Queen Gertrude
consented, and Elisabeth was sent to Thüringia, at the age of four, to
grow up in the Wartburg Castle, famous as the scene of the "Battle of
the Minstrels" spoken of by Rudolf Steiner in volume IV of his series of
lectures published under the title *Karmic Relationships*.

Elisabeth was a king's daughter, and she was beautiful. She awoke
love in those around her. Sophia, the wife of Landgrave Hermann
of Thüringia, who helped make the court at the Wartburg such an
extraordinary place for the Grail minstrel singers, took the child to
heart as an adoptive mother, alongside the many children of her own.

Growing up, Elisabeth felt herself drawn more and more to John
the Evangelist, under whose protection she placed herself. When
she was nine years old, the oldest son, Hermann, to whom she was
betrothed, died (in the year 1216). Landgrave Hermann's second son,
Ludwig, who was seven years older than Elisabeth, called her "my
little sister" and she replied "my dear brother." A strong bond grew
between them. They married in 1221, when Elisabeth was fourteen
years old. In the spring of 1222 Elisabeth gave birth to their first child,
a son, whom they named Hermann in honor of his grandfather, who
had died in 1217. Young Elisabeth was an extraordinary person. She
was not only beautiful but also full of magnanimity and warmth. We

can gather an impression of her qualities through the miracles that were recorded in the written accounts of her life.

The Miracles

One day, when Elisabeth went down from the Wartburg to give food to the poor, as was her custom, Ludwig came on his horse and wanted to know what she was carrying beneath her cloak. Pulling the cloak aside, he beheld roses. (This is the so-called "roses miracle" associated with Elisabeth.)

The St. Elisabeth Altar in the Elisabeth Church in Marburg, Germany, depicts a similar miracle that occurred when Elisabeth met a freezing beggar while on the way to join her husband at a banquet. She gave the beggar her cloak, which was richly embroidered to depict her rank at court. Because it would have been improper to arrive at the banquet without such a cloak, an angel replaced the one she had given away. (This is referred to as the "miracle of the cloak.")

Another time, to the horror of the entire court, Elisabeth took care of a leper, washing him, and then putting him to bed in the royal marital bed. Ludwig, upon returning to the Wartburg castle, heard about the leper and went to Elisabeth. Approaching the bed, he pulled back the covers. However, instead of a leper, he beheld the Crucified One. He recognized, then, that Christ is in every suffering human being.

So great was Elisabeth's fame on account of her holiness and piety that she received a visit from the emperor, Friedrich II, who was deeply moved by this encounter with Elisabeth—so much so, that after the death of Ludwig, the emperor (who had been a good friend of Ludwig's), requested Elisabeth's hand in marriage. However, she refused. Ludwig had died on the crusade that he joined in the summer of 1227, leaving Elisabeth behind, pregnant with their third child. Although marriage to the emperor would have offered her worldly security after Ludwig's death and no doubt would have afforded her many opportunities to help the poor, Elisabeth had a marriage of a different sort living in her heart. It was the mantle of St. Francis that marked the true glory of her crowning.

The Mantle of St. Francis

An unknown Franciscan monk, who was a contemporary of Elisabeth, reported that the fame of her piety had spread to Rome. St. Francis himself heard of her there from his friend and supporter, Cardinal Ugolino, who said to Francis:

> Dear brother, I would like you to leave something to this daughter in Germany, Elisabeth, who is filled with the same spirit as you, so that she can participate in your spirit, just as was the case with the disciple of the prophet Elijah, Elisha, when he picked up the mantle of the ascended prophet.[88]
>
> Elijah went up by a whirlwind into heaven. And Elisha saw it…and he took up the mantle of Elijah that had fallen from him…Now when the sons of the prophets who were at Jericho saw him over against them, they said, "The spirit of Elijah rests on Elisha." (II Kings 2:11–15)

Francis of Assisi took Cardinal Ugolino seriously and sent his mantle to Elisabeth, and according to tradition Elisabeth was buried wearing this mantle. It is as if Cardinal Ugolino sensed the great mystery connecting Francis of Assisi and Elisabeth of Thüringia, which Rudolf Steiner spoke of, indicating that they each had received an imprint of the *astral body* of Christ.[89] That they were both recipients of an imprint of Christ's astral body reveals the profound mystery underlying the extraordinary qualities of St. Francis and St. Elisabeth.

After the death of her husband Ludwig on September 11, 1227, followed shortly thereafter by the birth of her third child on September 29, Elisabeth was forced—by circumstances surrounding her pious behavior—to leave the castle. Elisabeth had refused to eat the food at court until she was sure that it had not been forcibly extorted from the peasants by Prince Heinrich, Ludwig's brother, who had become ruler after the death of Ludwig. Heinrich wanted Elisabeth's zeal for justice to be tamed and so did not give her the financial support to maintain her life at court. She was forced to leave the court and all the expectations of her social position there, something she had endured as an inner conflict when confronted by the suffering and poverty outside of the court.

In light of these harsh circumstances, it is remarkable that Elisabeth did not accept the emperor's proposal to become empress of Europe. She chose instead accommodations in Eisenach, for a time, in a converted pig sty. In 1228, she left Thüringia and went to Marburg, where she founded a hospital and a chapel, which she dedicated to St. Francis. One day after she had bathed some impoverished people in her Marburg hospital, put them to bed and covered them, she was heard to say, "How happy for us to have the honor of bathing our Lord and covering him."

Elisabeth had chosen to serve the poor, and entered the Third Order of St. Francis in 1229, renouncing her children and her relatives. She took her vows on Good Friday, 1229, in the presence of her father-confessor, the Franciscan Konrad of Marburg, who then began to exert a strong (and unhealthy) influence upon Elisabeth. He lived a life of extreme poverty and treated himself harshly. As Elisabeth's father-confessor, he demanded total obedience from her and treated her harshly, laying extraordinary demands of self-sacrifice upon her. Through Konrad of Marburg she was drawn into unnecessary suffering, which may have contributed to her early death at the age of twenty-four.

Against Konrad of Marburg's wish, she followed him to Marburg where, under his direction, she took up the task of caring for the sick in the hospital in Marburg that she had founded. Everyone, including Konrad, was astonished at her healing knowledge and powers. It was primarily on account of her deeds of love for the sick and poor who came to her hospital that Elisabeth was pronounced a saint at Pentecost 1235, just four years after her death in 1231. Konrad of Marburg reported Elisabeth's death as follows:

> When I asked her what she wanted to do with her property and belongings, she answered that everything regarded as hers belonged to the poor.... Then she spoke...quite especially about the raising of Lazarus and how the Lord wept at his raising. And when the monks and nuns were moved to tears through these words, she said: "You daughters of Jerusalem, do not weep on my account, but for your own sake." Then she ceased talking and the sweetest tones were heard in her throat

without her moving her lips. And when those present asked
what this was, she asked if they also had not heard the singing
voices. From now on—it was dusk—she lay filled with heavenly
joy and with signs of the deepest rapture until the cock's first
crow.... She died on November 17.[90]

St. Elisabeth's Birth—Battle of the Minstrels

St. Elisabeth died, therefore, on November 17, 1231, at about
five A.M. in Marburg. Which brings us to the question, when was
she born? The answer to this question is connected with the "Battle
of the Minstrels" referred to by Rudolf Steiner. In fact, there were
two such battles at the Wartburg: in 1206 and in 1207. Among
the contestants who came together for this contest at the court of
Landgrave Hermann of Thüringia were Wolfram von Eschenbach,
Walther von der Vogelweide, and Heinrich von Ofterdingen. Wolfram
wrote part of his *Parzival* epic at the Wartburg, completing this work,
it is believed, between 1209 and 1215. He enjoyed the patronage of
Landgrave Hermann, who later furnished him with a French manuscript about the knight Count William of Orange, which formed
the subject matter for Wolfram's epic poem *Willehelm* that he commenced after completing *Parzival*, but never finished, only nine of the
ten books of *Willehelm* being extant.

The singing contest known as the "Battle of the Minstrels" is
the subject of the second act of Richard Wagner's opera *Tannhäuser*.
However, it is doubtful that Wolfram von Eschenbach ever met
Tannhäuser, who was a real person, as Wagner portrays.

The first Battle of the Minstrels took place at the Wartburg in July
of 1206. On account of dramatic turn of events, the minstrels were
summoned by Landgrave Hermann to return for a second contest one
year later, which took place on July 7, 1207, again at the Wartburg.
What was the reason for this?

At the first Battle of the Minstrels in 1206, Landgrave Hermann,
the great patron of the minstrels, had summoned the minstrels to
sing of the knightly virtues. Naturally, he hoped that they would
sing about *his* knightly virtues. Heinrich von Ofterdingen opened the
festival with a song of praise to Landgrave Hermann, as an act of

politeness. Then, however, he sang of Duke Leopold VI of Austria as the most virtuous knight, whom he compared with the Sun. This was a public insult to Landgrave Hermann, and the other minstrels came to the Landgrave's defense. Walther von der Vogelweide sang in rage that the hangman should be summoned on account of Heinrich von Ofterdingen. Thus the hangman was summoned, and it was decided that the loser at this singing contest should be hung. When it became clear that Heinrich von Ofterdingen had lost, the hangman approached him. At this moment the Landgrave's wife, Sophia, intervened to save Heinrich's life, even though Heinrich had sung about Klingsor from Hungary as the greatest singer of all, who would sing the praises of Duke Leopold of Austria.

Sophia pleaded for the contest to be postponed, to allow Heinrich to summon Klingsor from Hungary, and then to continue. This was agreed, and the date of July 7 the next year, 1207, was set as the day upon which all the minstrels, including Heinrich in the company of Klingsor, should return for the continuation of the Battle of the Minstrels.

It was on July 7, 1207, that the singing contest between Wolfram von Eschenbach and Klingsor, referred to by Rudolf Steiner, took place:[91]

A new Battle of the Wartburg ensues and Klingsor enters the lists for Heinrich von Ofterdingen. Klingsor himself sings against the others, but it is quite evident that he is not battling alone. He causes spiritual beings to battle for him. For instance, in order to do so, he makes a youth become possessed by a spiritual being—and then compels the youth to sing in his place. He calls still stronger spiritual forces into play in the Wartburg.

Over and against all that came from Klingsor's side stands Wolfram von Eschenbach. One of Klingsor's practices is to make one of his spiritual beings put Wolfram to the test, as to whether he is really a learned man. Even so Klingsor finds himself driven into a corner by Wolfram. In effect, Wolfram von Eschenbach, observing that some spiritual influence is at work, sings of the Holy Communion, the Transubstantiation, the Presence of Christ in the Eucharist, and the spirit is obliged to depart, for he cannot bear it.[92]

St. Elisabeth's Birth

There are other interesting things referred to by Rudolf Steiner in this lecture, which are directly relevant to our theme of the development of a new star wisdom. The question with which we are occupied at present is: When was St. Elisabeth born?

> The date July 7, 1207, was not only that of the Battle of the Minstrels. That very night Klingsor announced that a daughter had just been born to his lord, King Andreas II of Hungary. She will be betrothed to the son of the Landgrave of Thüringia, and through her sanctity she will be a comfort to the whole of Christendom.[93]

Reading this, it would appear, as Erica von Dellingshausen concludes, that Elisabeth was born on July 7, 1207. Apparently, however, there exists a plaque in the chapel of the castle of Sárospatak where Elisabeth was born that states July 5, 1207, to have been the date of Elisabeth's birth. This information came to me through Christa von Cappelen who, having a great reverence for St. Elisabeth, had just been to Hungary and had seen the plaque. She communicated St. Elisabeth's date of birth to me as July 5, 1207, quite independently of any discussion about the Battle of the Wartburg on July 7, 1207. Reading Klingsor's announcement that "a daughter had just been born" would allow for a birth two days previously.

The results of all the chart comparisons (St. Elisabeth—Raphael, St. Elisabeth—Novalis, etc.) confirm July 5 rather than July 7, 1207, as the most likely date of birth. Although I have not been able to find written confirmation of Christa von Cappelen's communication, I now have no reason to doubt the authenticity of the two-day difference—which affects the positions of the Sun, the Moon, Mercury and Venus, also Mars, to a slight extent. As I am convinced that July 5, 1207 (rather than July 7, 1207) is the correct birth date of St. Elisabeth, this is the date I shall work with when it comes to a comparison of charts with those of other incarnations of this individuality: John the Baptist, Raphael, and Novalis. The first question, however, is: Was Elisabeth of Hungary indeed the "saintly nun" whom Rudolf Steiner had in mind when he referred to a previous incarnation of Novalis?

The "Saintly Nun" Hypothesis

When I say—speaking as a scientist and mathematician—that I know with *absolute certainty* that John the Baptist reincarnated as St. Elisabeth of Hungary and St. Elisabeth as Raphael and Raphael as Novalis, you, the reader are initially put in a position of suspending judgment for the time being and regarding the inclusion of St. Elisabeth (in this sequence given by Rudolf Steiner) as a hypothesis. There is nothing to be lost by regarding it as a hypothesis and considering the evidence.

Before presenting the evidence, it will be important to consider that I did not arrive at this hypothesis by way of intellectual speculation. To put it succinctly, it was through spiritual cognition that I discovered that St. Elisabeth reincarnated as Raphael. At the time I did not know of Rudolf Steiner's statement concerning "saintly nuns." When I read it, it served as direct confirmation of what I already knew. I knew it—and know it—with certainty, just as through being a mathematician and scientist, I can know that the Pythagorean theorem is true, or that there are stars in the heavens. This process of direct knowing is called *gnosis*. In line with the perspective of gnosis, let us now ask the question: How can we trust what is found or communicated by way of spiritual cognition or gnosis? What is the evidence?

In this case of St. Elisabeth, there is the evidence of astrological reincarnation research. And second, there is "biographical evidence" or "karmic evidence," whereby a *signature* is given from incarnation to incarnation. Let us begin with this latter kind of evidence. While realizing that the evidence is in no way conclusive, it nevertheless offers a starting point for our considerations.

St. Elisabeth—John the Baptist

Jesus Christ said of John the Baptist: "There has risen no one greater than John the Baptist" (*Matthew* 11:11). This individuality is a great individuality, one very close to Christ. The fact that St. Elisabeth of Thüringia received the astral body of Christ points to a close connection with him. She was undoubtedly a great individual, singled out by Christ together with St. Francis of Assisi and also St. Thomas Aquinas (according to Rudolf Steiner), for this special divine grace.

John the Baptist lived a short life, dying at the age of 31 years, 7 months. The "signature" of a short life we find also in the case of Raphael, who died at the age of 37 years, 9 days, and in the case of Novalis, who died at 28 years, 11 months of age. At her death St. Elisabeth was 24 years, 4 months old.

John the Baptist died at the hands of the "black magical" stream represented by Herod Antipas, Herodias, and Salome, the daughter of Herodias (Matthew 14:1-11). The birth of St. Elisabeth was announced by the black magician Klingsor.

The name of John the Baptist's mother was Elisabeth, which means "house of Elijah," and she indeed "housed" the reincarnated Elijah in her womb. St. Elisabeth of Thüringia "housed" the reincarnated Elijah in her very being.

According to Rudolf Steiner in his "last address"—his last lecture, held on September 28, 1924—he referred to the raising of Lazarus, saying that the individuality of John the Baptist, who had been beheaded 1½ years prior to the raising of Lazarus, descended from spiritual realms to *unite* spiritually with Lazarus.[94] From this point in time onwards it is possible to speak of Lazarus–John. Let us recall the tender words of St. Elisabeth on her deathbed referring to the raising of Lazarus. According to the account reported by Konrad of Marburg concerning her death: "She spoke quite especially about the *raising of Lazarus*, and of how the Lord wept at his raising"—and of how due to the *living quality* of her words "the monks and priests were brought to tears." Further Rudolf Steiner spoke of Lazarus–John as the writer of the Gospel of St. John, i.e., John the Evangelist.[95] Let us recall that, already as a young girl, Elisabeth of Thüringia placed herself under the protection of John the Evangelist.

In light of Rudolf Steiner's indication on the one hand of the profound spiritual relationship between John the Baptist and John the Evangelist, and on the other hand that John the Baptist reincarnated as Novalis, the words of Novalis's friend Ludwig Tieck are most striking:

> Novalis was tall, slender, and of noble proportions. He wore his light brown hair in locks that fell to his shoulders. His brown eyes were bright and luminous and his complexion, especially

above his intelligent brow, was almost transparent.... The set
and expression of his features came very close to those of St.
John the Evangelist, as shown on the fine, great tablet from the
brush of Albrecht Dürer, which is preserved at Nuremberg and
Munich.... Without the slightest vanity or intellectual arro-
gance, a stranger to all affectation and hypocrisy, he was a man
in the truest sense, the finest and purest embodiment of a lofty,
immortal spirit.[96]

St. Elisabeth—Raphael

Let us now turn our attention to consider Raphael, the incarnation of the John the Baptist individuality preceding that of Novalis
and following that of St. Elisabeth. What "karmic biographical" connections do we find between St. Elisabeth and Raphael?

First, the name Raphael was not a mere coincidence. Rather, it
was given to the child by his parents through inspiration. For there
is a deep connection between the Archangel Raphael and the painter
Raphael, indeed, with the individuality: Elijah—John the Baptist—St.
Elisabeth—Raphael—Novalis. The Archangel Raphael is associated
with Mercury, with breathing, and with healing.[97] Against this background we behold the spiritual source of the healing impulse and healing powers of St. Elisabeth. Raphael's majestic works of art, particularly his Madonna paintings, embody a healing power. Most of the
sequence of fifteen Madonna paintings that were indicated by Rudolf
Steiner for contemplation as a "healing sequence" are by Raphael.[98]
In this light it is possible to consider that the name "Raphael" given to
him by his parents expresses an inner relationship of this individuality
to the Archangel Raphael.

Second, Raphael's contemporaries saw him as a saintly person.
In other words, his previous incarnation as a "saintly nun" shone
through him, so much so that the Pope was thinking of pronouncing
him a saint (after his death), as the following anecdote relating to "La
Fornarina" shows:

The *original* for the beautiful women Raphael painted in
his Roman period was La Fornarina. Raphael surprised her
while she was wading in a river and "since he was desirous

of every kind of beauty, finding her beautiful above all things, he straightway became enamored of her and fixed all his thoughts upon her, nor had he any peace until he had made her his." La Fornarina is present in spirit in all his works. During his last illness she was constantly at his bedside. In his will he left her a share of his fortune so she might "live an honest woman." When the funeral procession wound its way through the streets of Rome, La Fornarina escaped from the people who were guarding her and flung herself upon the coffin, and allowed herself to be drawn away only when she was told that the Pope was thinking of beatifying Raphael and that therefore there must be no more talk of his earthly affections.[99]

Raphael had a definite karma with the Roman Catholic Church, which could be seen as an echo of his incarnation as a saintly nun, St. Elisabeth, who had become a nun in the Third Order (Franciscan) at the age of twenty-one years. His karma with the Church is evident from the considerations to elect him as a cardinal at the Vatican without taking holy orders:

> With his religious temperament, his good heart, and his excellent relations with the papal court, Raphael hoped at one point in his life, not without justification, to be named a cardinal. This was in fact suggested by [Pope] Leo X's advisers, and by [Cardinal] Bibbiena in particular. Vasari writes, "Having for so many years served the [papal] court and seeing Leo X in his debt for a goodly sum, he was encouraged to believe that, as a just reward for his labors and his many virtues, the Pope would give him a red hat." And he would most probably have become a cardinal had he not died at the early age of thirty-seven. This is not as surprising as it sounds if we remember that at the time even a citizen who had never taken holy orders of any kind could be named cardinal by the Pope. It was required, however, that he should be unmarried and to be well known for piety and good moral character.[100]

Recalling the spiritual connection between St. Elisabeth of Hungary and St. Francis of Assisi, it is interesting that St. Elisabeth—

reincarnating as Raphael—chose to be born at Urbino in Umbria, the region of Italy associated with St. Francis. Raphael we recall "was born and died on the same day" Good Friday, and that Good Friday was the holy day that St. Elizabeth had chosen for the taking of the Franciscan vows when she became a nun.

The themes of Raphael's paintings—he is famous for his numerous portrayals of the Madonna, including the "Sistine Madonna"—can be experienced as most tender and intimate, as though painted in *remembrance* of his incarnation as John the Baptist. The painting of the "Transfiguration" was his last painting and, although not completed was placed at his bedside when he died—as a symbol, perhaps, of an ongoing prophecy, an indicator of a "golden thread" of spiritual intention to return as a living herald for the story of Christ that was not yet finished. The painting of the "Transfiguration," understood from this future perspective yet to come, serves humanity as both proclamation and promise of the coming of a *transfigured* Christ with *Moses* holding the tablet of the divine law on Christ's right and *Elijah* bearing witness to the Christ on Christ's left (see front cover image). As an artist, Raphael had served as a faithful witness to Christ and his "death was felt by all like the death of a God, and the end of a myth."[101]

St. Elisabeth—Novalis

The theme of Mary returned in Raphael's subsequent incarnation as Novalis, since Mary was a source of inspiration for Novalis in his poetry. Novalis's reverence for Mary comes to expression especially in the two final hymns of *Hymns to the Night*, singing forth the sublime nature of the Eternal Feminine—so beautifully rendered in the numerous paintings of Raphael's Madonnas. There are other themes in the story of the life of Novalis, which are more directly reminiscent of the life of St. Elisabeth.

"Scion of an ancient family of Lower Saxony, going back to the Hohenstaufen dynasty and the War of the Minstrels at Wartburg castle,"[102] Novalis grew up in the cold, austere atmosphere of an old and rambling family estate that had once housed a nunnery. The dry and austere atmosphere was heightened by the puritanical practices of his Pietist father, who tyrannized the family with strict requirements around daily routine, devotions, bible lessons, etc. This is entirely reminiscent of St. Elisabeth's father-confessor, "the grand inquisitor of Germany, Konrad of Marburg.... He drove her to the most severe penitential exercises in such bad physical circumstances, that she soon came down with tuberculosis...,"[103] as did Novalis shortly before his twenty-ninth year.

The frugal conditions of Novalis's early childhood awakened in him a thirst for knowledge, so that later when he was allowed as a youth to live with a worldly, sophisticated uncle, much like a sign of *pentimento* in a painting—the "bleed through" of an under-painting peeking through from an earlier time—Novalis requested an Italian grammar and a catalogue of Italian books. Harkening back to the cultural sophistication of Raphael's birth place in Urbino, there in the estate of his uncle was a grand library, referred to by Novalis as the "Grand Cross," and a spirit of the Age of Enlightenment and "courtly absolutism" prevailed. We recall the high ideals and moral code set forth as a standard of behavior for the true "gentleman" of the Renaissance, practiced in Urbino, Italy, during Raphael's youth— a time when the ideal of gentlemanly behavior was to be "a man of honor and decency."[104]

Seeming as an echo from the time of the Renaissance, Frederick Wilhelm von Hardenberg, the elder brother of Novalis's pious father, was a "Lord of the German Order of Knighthood and Knight Master of its Commandery of Saxony."[105] Naturally, his estate was a meeting place of aristocrats. Recalling now the earlier "knightly" code held forth in Thüringia, Germany, when the Battle of the Minstrels was fought under the banner of knighthood and hinged upon the question about the knightly virtues, it seems important to consider that this was also the occasion of the proclamation of the birth of a daughter to the King and Queen of Hungary, who was to be bequeathed in marriage to the eldest son of Landgrave Hermann I of Thüringia, would come to live at the Wartburg castle, and would be as a consolation and a healing to all of Christendom. Thus Elisabeth's birth was indelibly linked with the question of the perfection of the human being, the yet unfinished story of becoming....

As an answering accord to this "golden thread" of connection Novalis wrote: "If the idea is not put into thought patterns, the ideal cannot be born!" Thus in the novel *Heinrich von Ofterdingen*, Novalis tells the story of the *"becoming* of the total human being" giving birth to the "idea of the ideal." Embedded in the novel is the fairy tale "Eros and Fable"—which was written toward the end of Novalis's life—in which he describes a time in the future when *"the evil forces of the world are overcome and a new earth comes into existence. Human beings have learned how to act, life has become a healing and ordering experience...."* This is a description of Novalis's dream for the future of human beings: *"Humankind representing the higher purposes of our planet."*[106]

From the memoirs of Henrik Steffens, one of Novalis's contemporaries, there is yet another glimpse of *pentimento*: "On first impression his appearance recalled those devout Christians whose nature is simplicity itself.... His deep-seated eyes, above all, carried an ethereal fire.... Few men have left such a deep and lasting impression on me.... Novalis always seemed to speak as well as to write from a deep sense of the past."[107]

Could it be that it was this deep sense of the past that led Novalis to reincarnate in the region of Thüringia? "Thüringia is seen by many to be the middle—the heart of Europe—it not only has the task of mediating between East and West, but also, is a place where new impulses arise such as, for example, the hospital founded by St. Elisabeth."[108]

Novalis, who was born in Thüringia and whose poetry encompassed a passionate call for the "arising of a new earth," was on a mineralogical excursion through Thüringia in mid-June, the final summer of his life. Again the theme of *pentimento* was indicated by friends whom he visited at the foot of Kyffhäuser mountain in Thüringia, who recounted that Novalis had been studying the *Chronicles of Thüringia*, as well as reading Johannes Roth's *Life of St. Elisabeth* and the *The History of Emperor Frederick II*, the emperor who had proposed marriage to St. Elisabeth and who later served as coffin bearer at her death. Upon returning home Novalis prepared a petition applying for a post as district justiciary in Thüringia, which was awarded him shortly before his death.[109]

Novalis's Novel—*Heinrich von Ofterdingen* (*Henry of Ofterdingen*)

In Novalis's great novel, *Heinrich von Ofterdingen*, a clear, inner *remembrance* back to the Wartburg and the Battle of the Minstrels is revealed that, against the background of Novalis's last-but-one incarnation as St. Elisabeth of Thüringia, is illuminating. In his novel Novalis depicts Klingsor ("Klingsohr" in the spelling of Novalis), the magician summoned to the Wartburg by Heinrich von Ofterdingen, as benevolent. Heinrich travels to Augsburg, where he meets Klingsor, who becomes his poetic mentor. However, in Rudolf Steiner's account of the Battle of the Minstrels referred to above, Klingsor was a not-so-benevolent magician who called forth evil spirits to sing through the young Heinrich von Ofterdingen on his behalf and thus influence the outcome of the battle by way of distorting the truth. Now, in Novalis's novel *Heinrich von Ofterdingen* we experience an evolved sense for the truth. Essentially Novalis redeems Klingsor by endeavoring to set things right in the realm of magic, calling for *sacred magic*—to call forth the help of the heavenly hierarchies to inform one's poetic songs. This holy magic of the word is manifested in the quality of the sacred in much of Novalis's poetry and is especially evident in his *Sacred Songs*. Novalis, who felt himself to be a servant of the Word or Logos, wrote: "Language is the dynamic element of the spiritual realm."[110] He was conscious of carrying over the "dynamic element of the spiritual realm" into the earthly realm through his poetry and writing, in this sense calling forth the spiritual powers on high to intervene in the earthly realm—a kind of *sacred magic*.

Returning to the unfolding story in his novel *Heinrich von Ofterdingen*, Klingsor has a beautiful daughter, Mathilda, whom Novalis describes as the "visible spirit of song" and with whom Heinrich falls in love. Here we recall the beatific radiance and the sound of singing coming from the throat of St. Elisabeth when she died. Mathilda tells Heinrich "about Hungary, where her father often stayed." Later, Klingsor asks the lovely Mathilda, "Would you like to be Heinrich's inseparable companion?" Then Klingsor tells Heinrich all about poetry and the mission of the poet, whereby Novalis drew heavily upon Goethe's views concerning order and discipline in poetry. There one could say that the redeemed Klingsor offers a kind of "portrait" of the poet Goethe.

With the betrothal of Heinrich von Ofterdingen and Mathilda, Klingsor recounts a strange and complex fairy tale, "Eros and Fable," that was inspired by Goethe's "Fairy Tale" published in 1795. Its theme is one of struggle between cold, calculating reason and the spontaneity of life and poetry. Sophie (Wisdom) upholds the latter. She proclaims that, "The great mystery is revealed to all," and later says, referring to the Divine Mother, "Mother is among us, her presence will bless us forever." The closing words to "Eros and Fable" are:

> The kingdom of eternity is founded,
> In love and peace all strife is concluded,
> The long dream of pains is now past,
> Sophie is eternal Priestess of all hearts.[111]

This concludes the first part of the novel (entitled "Expectation"). The second part ("Fulfillment") remained only a fragment. The key to *Heinrich von Ofterdingen* is that it is not only autobiographical but also "karmic biographical," albeit in a rather complex form. Let us recall that the life of St. Elisabeth of Hungary began with Klingsor's announcement of her birth at the Battle of the Minstrels at the Wartburg on July 7, 1207—her birth having taken place just two days prior to this—and that Klingsor announced prophetically her coming to the Wartburg. Somehow there was a spiritual link between the incarnation/birth of St. Elisabeth and the Battle of the Minstrels, at which Klingsor and Heinrich von Ofterdingen played key roles. Somewhere within the soul of Elisabeth she sought to bring to resolution the Battle of the Wartburg with which she was spiritually connected. The resolution is depicted in Novalis's novel *Heinrich von Ofterdingen*, where Heinrich is the hero, upon whom aspects of Novalis himself are projected, and where Klingsor appears in a benevolent form. Novalis wrestled with Goethe's "classical" conception of poetry, to give birth to romantic poetry. Goethe was thus a kind of poetic mentor of Novalis, just as (in the novel) Klingsor was mentor of the budding poet, Heinrich von Ofterdingen.

Apart from being a poet, Novalis worked as a mine inspector in Weissenfels. An autobiographical note emerges in the novel where Heinrich von Ofterdingen meets a miner, who initiates him into the Earth mysteries: "Mining must bear God's blessing...man is the Lord of the Earth...as though she were his bride."[112] Another autobiographical theme is the relationship between Heinrich and Mathilda, mirroring Novalis's relationship with his fiancée, Sophie von Kühn. Tragically, Sophie died. In dying, however, she became Novalis's muse. He formed a "mystical" relationship with Sophie, and his real birth as a poet took place through her, following her death. Similarly, in the novel, Mathilda dies. Yet she becomes, in this way, the inspiration for Heinrich's birth as a poet.

A karmic biographical element shines through the novel in that Heinrich's home town is Eisenach, next to the Wartburg, and the Landgravine Sophia of the Wartburg is Heinrich's "godmother." Let us remember that, when four-year old Elisabeth arrived at the Wartburg, Landgravine Sophia adopted her into her family as a "daughter" alongside her many children. Later St. Elisabeth left the Wartburg/Eisenach region and went to Marburg. In the novel Heinrich von Ofterdingen left Eisenach and went to Augsburg. Could it be that Novalis inwardly—through his gift of inspiration—heard the word "Marburg" yet substituted "Augsburg" in its place? At any rate, the fact is that Novalis reincarnated in Thüringia, the region associated with St. Elisabeth.

Novalis's inspiration for his novel *Heinrich von Ofterdingen* was, I believe, a karmic memory, albeit on a subconscious level, of his last-but-one incarnation as St. Elisabeth at the Wartburg, the scene of the Battle of the Minstrels. And autobiographical themes from his life as Novalis were interwoven with this karmic memory.

Novalis's Essay—"Christendom or Europe"
Although by some accounts Novalis's writing was thought to be typical of the Romantic literature at the end of the eighteenth century, what sets it apart is that in it can be found, "A depth and certainty of Christian experience"—a sense of immediacy and intimacy, reflecting the sound and substance of the deeply personal—that was otherwise

"rare among his contemporaries."[113] It is against the background of Novalis's incarnation as St. Elisabeth—whose birth was announced among the Grail minstrel singers at the singing contest referred to as the "Battle of the Minstrels" as one who "through her sanctity would bring comfort and healing to all of Christendom"—that now we can recognize how Novalis came to write the following lines:

> Once there were fine, resplendent times when Europe was a Christian land, when one Christendom occupied this humanly constituted continent. One great common interest united the remotest provinces of this broad spiritual realm. Without great worldly possessions, one Head guided and unified the great political forces. A numerous guild, to which everyone had access, stood directly beneath him, and carried out his behests and strove with zeal to confirm his beneficent power. Every member of this organization was universally honored, and if the common people sought comfort or help, protection or counsel from this member, and in return were happy to provide generously for his manifold needs, he also found protection, respect, and a hearing among the more powerful, and every-one cared for these chosen men, equipped with miraculous powers, as for children of Heaven whose presence and favor spread manifold blessing abroad. Childlike faith bound men to their pronouncements. How cheerfully every man could fulfill his earthly labors when, through the agency of these holy persons, a secure future was prepared for him and every misstep forgiven, when every discolored spot in life was obliterated by them and made clean. They were the experienced helmsmen upon the great unknown sea, in whose keeping one might disdain all storms and count on a sure attainment of the coast and a landing at the world of the true home.
>
> Before their words the wildest and most voracious propensities were obliged to yield respect and obedience. Peace proceeded from them. They preached solely love for the holy and wondrously beautiful Lady of Christendom, who, endowed with divine powers, was prepared to rescue any believer from the most dread perils. They told of celestial persons long since dead who, by virtue of adherence and loyalty to that Blessed

Mother and to her divine and benevolent Child, withstood the
temptation of the earthly world and achieved honors and had
now become protective and beneficent powers to their living
brethren, willing helpers in tribulation, intercessors for human
infirmities, and efficacious friends of mankind before the heavenly throne. With what serenity people used to depart from the
beautiful assemblies in the mysterious churches, which were
adorned with cheering pictures, filled with sweet fragrances,
and animated by holy and exalting music. Therein the conse-
crated remains of former God-fearing men were gratefully preserved in precious reliquaries. And through them was manifest
the divine goodness and omnipotence, the powerful beneficence
of these happy saints, in splendid wonders and signs. In this
way loving souls preserve locks of hair or bits of writing of
their departed loved ones and feed the sweet flame thereby until
reuniting death. With heartfelt care people used to gather from
everywhere whatever had belonged to these beloved souls. And
each man considered himself fortunate who was able to procure,
or so much as touch, such a consoling relic. Now and again the
heavenly grace seemed to have descended especially upon some
strange picture or upon a grave. Thither streamed people then
from all regions with lovely gifts, and carried away heavenly
gifts in return: peace of soul and health of body.

Assiduously this powerful peace-creating organization sought
to make all men sharers in this beautiful faith and sent their col-
leagues into all parts of the world to proclaim everywhere the
Gospel of Life and to make the Kingdom of Heaven the only
kingdom on this Earth. With good cause the wise Head of the
Church countered insolent excrescences of human talents at the
expense of the sacred sense, as well as untimely, dangerous discoveries in the area of knowledge. Thus he prevented bold think-
ers from asserting publicly that the Earth was an insignificant
planet, for he realized that humans, together with respect for
their dwelling place and their earthly homeland, would also lose
respect for their heavenly home and for their race, would prefer
circumscribed knowledge to infinite faith, and would become
accustomed to scorning everything great and worthy of wonder
and look upon these as dead legalisms. At his court assembled

all the clever and revered men in Europe. All treasures flowed thither, Jerusalem destroyed had avenged itself, and Rome itself was Jerusalem, the holy residence of divine government on Earth. Princes laid their disputes before the father of Christendom, willingly laid their crowns and their splendor at his feet. Indeed, they deemed it a glory to conclude the evening of their lives as members of that high guild in godly contemplation within solitary cloister walls. How beneficial this regimen, this arrangement was, how appropriate to the inner nature of man, was shown by the mighty upsurge of all the other human powers, the harmonious development of all capacities, the tremendous height to which individual men attained in all departments of knowledge of life and of the arts, and by the universally flourishing traffic in spiritual and earthly wares within the boundaries of Europe and outward to the most distant Indies.

Such were the fine essential characteristics of the truly Catholic or truly Christian times....

Applied, vitalized Christianity was the old Catholic faith, the last of these forms. Its omnipresence in life, its love of art, its profound humanity, the inviolability of its marriages, its communicativeness benevolent to man, its joy in poverty, obedience, and loyalty, render it unmistakable as genuine religion...[114]

It took twenty-five years for Novalis's essay "Christendom or Europe" to appear after his death. Yet, for Novalis who believed that "language is the dynamic element of the spiritual realm,"[115] it was important that the ideas be born, and indeed the essay "Christendom or Europe" does serve as a "golden thread" of spiritual intention linking Novalis to his earlier incarnation as St. Elisabeth of Hungary. His words and thoughts have the sounding substance of a karmic memory coming from his incarnation as the "saintly nun" (Rudolf Steiner). Taken together with Rudolf Steiner's indication of Novalis's previous incarnation as Raphael, let us now take a look at the horoscope comparisons between the incarnations: St. Elisabeth—Raphael—Novalis.

9

St. Elizabeth – Raphael – Novalis:
Horoscope Comparisons

I N PRESENTING THE FOREGOING, I have revealed something of the biographical *signature* of John the Baptist, Raphael, and Novalis in relation to the incarnation of St. Elisabeth of Thüringia—which was presented for the reader as a hypothesis for the solving of the mystery of the identity of the "saintly nun" referred to by Rudolf Steiner in relation to Novalis as one of the previous incarnations of this individuality. In the following we shall come to the findings of astrological reincarnation research regarding the sequence of incarnations: St. Elisabeth of Thuringia—Raphael—Novalis, in the quest to discover what further evidence exists (if any) in support of this hypothesis.

St. Elisabeth – Raphael

Uranus

First of all, let us consider the relationship between the horoscope at the *death* of St. Elisabeth and that at the *birth* of Raphael (see comparison chart in appendix 2). Recalling the apparent significance pointed out in chapter 1 of the slow moving outer planets—Uranus, Neptune, and Pluto—when a complete number of orbits takes place between incarnations, the first thing to notice in comparing the death of St. Elisabeth with the birth of Raphael is the *exact* alignment of *geocentric Uranus* in the comparison chart. It is interesting to note, further, that the Uranus alignment at 2° Sagittarius indicates an alignment with the *Galactic Center:*[116]

> *St. Elisabeth Death g-Uranus (2♐10)—conjunct—*
> *Raphael Birth g-Uranus (2♐21)*

This means that Uranus made a complete number of revolutions of the sidereal zodiac between the death of St. Elisabeth in 1231 and the birth of Raphael in 1483. In this period of 252 years, Uranus made three orbits of the zodiac, since the orbital period of Uranus is 84 years (3 x 84 = 252). This is exactly the same number of orbits as between the death of Raphael in 1520 and the birth of Novalis in 1772, i.e., 252 years—again with the alignment of geocentric Uranus from death to rebirth being exact.

St. Elisabeth Death—three Uranus orbits of the sidereal zodiac—Raphael Birth
Raphael Death—three Uranus orbits of the sidereal zodiac—Novalis Birth

Regarding St. Elisabeth, Raphael, and Novalis as three successive incarnations of the same individuality, we see a strong relationship to the planet *Uranus*, with this individuality choosing to reincarnate after "spiritually observing" Uranus to make three orbits of the zodiac, then to return to alignment with its position at death in the preceding incarnation. What is meant here by "spiritually observing" is the soul's mode of cognition, while indwelling the cosmic realms between death and a new birth, of the surrounding cosmos in which the human soul is immersed at that time. Of course, *it would be premature—on the basis of fulfillment in two instances—to speak of a "law" here relating to the planet Uranus.* Nevertheless, it is worthy of further investigation.

Mars / Sun Opposition:

Another striking alignment between the death of St. Elisabeth and the birth of Raphael is the geocentric *Sun/Mars* opposition:

St. Elisabeth Death Sun (17♏47)—opposite—
Raphael Birth g-Mars (18♉05)

Here we discover that the position of *Mars* geocentrically at Raphael's *birth* aligned almost exactly *opposite* the *Sun's* position at St. Elisabeth's *death*.

In comparing the *death* of St. Elisabeth with the *death* of Raphael (see comparison chart in appendix 2), we see the *Sun/Mars* opposition once again—this time in the *hermetic* chart:

> *St. Elisabeth Death h-Mars (6♎49)—opposite—*
> *Raphael Death Sun (7♈57)*

There are some other close alignments in the geocentric and heliocentric/hermetic chart comparisons between the death of St. Elisabeth and the birth of Raphael. However, neither the first nor the second "law" of astrological reincarnation is fulfilled. On the basis of years of experience it is possible to say that it is not unusual in astrological reincarnation research to find reincarnation examples in which neither of the two "laws" holds—one example being that of Raphael–Novalis.

Without looking at any further alignments between the birth and death horoscopes of St. Elisabeth and the epoch, birth, and death horoscopes of Raphael, let us now compare the charts of St. Elisabeth with those of Novalis. This is a unique opportunity to compare one incarnation (Novalis) with the last-but-one incarnation (St. Elisabeth). Let us note, though, that as the birth time of St. Elisabeth is not known, it is not possible to compute her conception/epoch horoscope, so her epoch chart has to be omitted from the comparisons.

St. Elisabeth Birth–Novalis Birth: Mars Alignments

First, comparing the birth of St. Elisabeth with the birth of Novalis (see appendix 2), we find an exact alignment of *geocentric Mars*, appearing as an opposition:

> *St. Elisabeth Birth g-Mars (13♍59)—opposite—*
> *Novalis Birth g-Mars (13♓20)*

Let us recall, as noted already in chapter 4, that in comparing the geocentric birth and death horoscopes of Novalis, we saw that there was an *alignment of the Sun in Pisces (12♓24) at Novalis's death with g-Mars (13♓20) at his birth*. The Sun at the death of Novalis and the geocentric position of Mars at his birth returned to the same zodiacal location *opposite* to that at which geocentric Mars had been at the birth of St. Elisabeth.

Now let us look at the *hermetic* chart comparison (see appendix 2), where there is another *Sun/Mars* alignment:

> *St. Elisabeth Birth h-Mars (21♎34)—opposite—*
> *Novalis Birth Sun (20♈53)*

Since the next comparison equation involves the Moon, it has to be borne in mind that in the case of the birth of St. Elisabeth the Moon's position is an average one for midday, as we do not know her time of birth. If she were born shortly after midnight, the Moon's longitude could be as much as seven degrees less than the value given in the comparison equation below, and if she were born shortly before midnight it could be as much as seven degrees more than the value below, which would then disqualify it from consideration as an alignment, as also would be the case if she were born in the early hours of the day. In other words, the following comparison equation holds only if she were born between about 2:30 AM and noon, since at 2:30 AM the Moon's longitude was 12♎15, within three degrees of 15♈08, and at noon the Moon's longitude was 17♎52, again within three degrees of 15♈08 (allowing an orb of about three degrees for such alignments):

> *St. Elisabeth Birth Moon (17♎52)—opposite—*
> *Novalis Birth Moon (15♈08)*

Taking noon as the average time of birth, when the actual time of birth is unknown, does not affect the planetary longitudes very much, except—as can be seen from the above example—in the case of the Moon (and also heliocentric/hermetic Mercury).[117] Moreover, not knowing the time of birth *precludes calculating the Ascendant*, which means, in turn, that *the conception/epoch horoscope also cannot be computed*.

St. Elisabeth Birth – Novalis Conception

Now let us compare the *birth* of St. Elisabeth with the *epoch/conception* of Novalis (see appendix 2). *Geocentrically* we find a close alignment of *Jupiter*, again an opposition:

> *St. Elisabeth Birth g-Jupiter (26♊08)—opposite—*
> *Novalis Epoch g-Jupiter (27♐55)*

Further geocentric alignments involving the Sun, Venus, and Pluto are in sidereal Cancer:

> *St. Elisabeth Birth Sun (5♋23)—conjunct—*
> *Novalis Epoch g-Venus (5♋48)*
> *St. Elisabeth Birth g-Pluto (23♋42)—conjunct —*
> *Novalis Epoch Sun (25♋24)*

Moon's Node Alignments

In the *hermetic* chart comparison (see appendix 2) there is an alignment of *Mars* and the *Ascending Moon Node* in sidereal Libra:

> *St. Elisabeth Birth h-Mars (21♎34)—conjunct—*
> *Novalis Epoch Ascending Moon Node (20♎44)*

The alignment of the Moon's Nodes with h-Mars is also significant in relation to the hermetic chart comparison between *St. Elizabeth* and *Raphael* (see appendix 2):

> *St. Elisabeth Birth h-Mars (21♎34)—conjunct —*
> *Raphael Birth Descending Moon Node (20♎36)*
> *St. Elisabeth Birth Ascending Moon Node (17♏20)—conjunct—*
> *Raphael Death h-Mars (15♏25)*

Alignments of the Moon's Nodes with Mars show up also *geocentrically*:

> *St. Elisabeth Death g-Mars (22♎18)—conjunct—*
> *Raphael Birth Descending Moon's Node (20♎36)*
> *St. Elisabeth Death g-Mars (22♎18)—conjunct—*
> *Raphael Death Descending Moon Node (24♎01)*

Returning now to the comparison chart between St. Elisabeth's birth and Novalis's death, another geocentric alignment is that of *g-Mars* with the *Descending Moon Node*, albeit this is a "loose conjunction" with an orb of just over three degrees (3°10')—taking three degrees as the usual orb allowed for an alignment:

> *St. Elisabeth Birth g-Mars (13♍59)—conjunct—*
> *Novalis Death Descending Moon Node (17♍09)*

The remarkable number of alignments involving the Moon's Nodes and Mars (heliocentric and geocentric) points to a definite relationship here, without going so far as to call it a "law." Astrological reincarnation research into further reincarnation examples is needed to clarify the exact nature of this relationship.

Mars / Sun Alignments

Geocentric alignments concerning *Mars* and the *Sun* in the comparison between St. Elisabeth and Novalis:

> *St. Elisabeth Birth g-Mars (13♍59)—opposite—*
> *Novalis Death Sun (12♓24)*
> *St. Elisabeth Death g-Mars (22♎18)—opposite—*
> *Novalis Birth Sun (20♈53)*

Conversely in the *hermetic* chart comparison a *Sun/Mars* alignment in sidereal Cancer shows up (see appendix 2):

> *St. Elisabeth Birth Sun (5♋23)—conjunct—*
> *Novalis Death h-Mars (2♋48)*

Taking these alignments together with the Mars/Mars and Mars/Sun alignments—Mars both heliocentric and geocentric—referred to earlier in this chapter, it is clear that the relationship between the Sun and Mars and also of Mars to itself (heliocentric and geocentric) is of special significance in the carrying over of destiny from one incarnation to later incarnations. Hopefully further astrological reincarnation research will be able to clarify the precise nature of the underlying relationship here.

Second "Law" Fulfillment

In this same *hermetic* chart comparison we are reminded of the *second "law"* (the orb of this alignment is almost 3½ degrees, which, although more than the usual three degrees, is still tolerable):

> *St. Elisabeth Birth ♄-Mercury (23♌59)—conjunct—*
> *Novalis Death ♄-Mercury (27♌28)*

And conversely there is also a *geocentric Mercury* alignment:

> *St. Elisabeth Death g-Mercury (9♏02)—opposite—*
> *Novalis Birth g-Mercury (10♉43)*

Further Alignments

We see that *geocentric Saturn* at Novalis's birth aligned with the *Moon* at St. Elisabeth's death (the Moon's position is fairly precise, as her approximate time of death is indicated):[118]

> *St. Elisabeth Death Moon (0♌36)—conjunct—*
> *Novalis Birth g-Saturn (1♌22)*

In the *hermetic* chart comparison there is a relationship between *Mars* and the *Moon's Node*, something that we have seen (above) over and over again:

> *St. Elisabeth Death ♄-Mars (6♎49)—conjunct—*
> *Novalis Birth Ascending Moon Node (7♎18)*

And this relationship shows up also in the geocentric chart comparison of St. Elisabeth's death chart with Novalis's *epoch/conception* horoscope:

> *St. Elisabeth Death g-Mars (22♎18)—conjunct—*
> *Novalis Epoch Ascending Moon Node (20♎44)*

These two alignments support the tentative conclusion drawn above concerning the relationship between the Moon's Node and Mars (heliocentric and geocentric) in astrological reincarnation research.

The following alignment of the *Sun* and *heliocentric Mercury* is interesting, as we shall see later:

> *St. Elisabeth Death ♄-Mercury (19♎30)—opposite—*
> *Novalis Birth Sun (20♈53)*

And there is also a geocentric *Sun/Jupiter* alignment between the death of St. Elisabeth and the conception/epoch of Novalis:

> *St. Elisabeth Death g-Jupiter (23♋12)—conjunct—*
> *Novalis Epoch Sun (25♋24)*

Echoes of the Second "Law"

In the hermetic chart comparison there is a fulfillment of the *second "law,"* although not between death and *birth* in the *next* incarnation but between death and *conception* in the *next-but-one* incarnation:

> *St. Elisabeth Death ♄-Venus (28♒07)—opposite—*
> *Novalis Epoch ♄-Mercury (0♍02)*

Moreover, here there is also an alignment of the *Ascending Moon Node* with *heliocentric Mercury*:

> *St. Elisabeth Death ♄-Mercury (19♎30)—conjunct—*
> *Novalis Epoch Ascending Moon Node (20♎44)*

In turn, this reminds us of an earlier alignment that we noted:

> *St. Elisabeth Death ♄-Mercury (19♎30)—conjunct—*
> *Raphael Birth Descending Moon Node (20♎36)*

St. Elisabeth Death—Novalis Death

Last, looking at connections between the *death* of St. Elisabeth and the *death* of Novalis, we see from the geocentric comparison chart a number of alignments involving the Moon's Nodes (see appendix 2). In order to present all these alignments in a more concise way, we shall now use the symbol † as an abbreviation for "Death" and A-Moon Node to denote "Ascending Moon Node" and D-Moon Node to denote "Descending Moon Node."

> † = Death
> A-Moon Node = Ascending Moon Node
> D-Moon Node = Descending Moon Node

Lunar Node Alignments

† St. Elisabeth A-Moon Node (27♎57)—conjunct—
  † Novalis g-Neptune (27♎08)
† St. Elisabeth D-Moon Node (27♈57)—conjunct—
  † Novalis g-Venus (28♈10)
† St. Elisabeth g-Saturn (17♓37)—conjunct—
  † Novalis A-Moon Node (17♓09)
† St. Elisabeth g-Pluto (15♍12)—conjunct—
  † Novalis D-Moon Node (17♍09)

The Second "Law"

In the hermetic death chart comparisons we are reminded of the *second "law"*:

† St. Elisabeth h-Venus (28♒07)—opposite—
  † Novalis h-Mercury (27♌28)

Having seen in the astrological reincarnation research presented in this book the second "law" relating to *h-Mercury/h-Venus align-ments* on several occasions extended beyond the strict formulation of it as applicable from death in one incarnation to birth in the next incarnation, it is obvious that it has a validity in an extended sense beyond this strict formulation. In addition we see:

† St. Elisabeth Moon (o♌36)—conjunct—
  † Novalis h-Saturn (o♌13)
† St. Elisabeth Moon (o♌36)—conjunct—
  † Novalis h-Venus (2♌44)

Jupiter Alignments

There is an interesting geocentric alignment between the two death horoscopes involving Jupiter and Saturn:

† St. Elisabeth g-Jupiter (23♋12)—conjunct—
  † Novalis g-Saturn (25♋41)

Most striking, however, is the alignment of *heliocentric Jupiter* in sidereal Cancer:

† *St. Elisabeth h-Jupiter (13♋29)—conjunct—*
  † *Novalis h-Jupiter (12♋40)*

This signifies exactly forty-eight orbits of heliocentric Jupiter around the sidereal zodiac between the death of St. Elisabeth in 1231 and the death of Novalis in 1801. The precise interval between the two deaths was 569.35 years, and since Jupiter's orbital period is 11.86 years, then 48 x 11.86 = 569.26. (Note that 11.86 years is a rounded figure, which accounts for the slight discrepancy here.)

From the above, it is evident that certain alignments recur fairly frequently:

> Heliocentric *Mercury/Venus* alignments (echoing the second "law");
> *h-Mars* and *g-Mars* alignments with the *Sun* and the *Moon's Nodes*;
> *Moon Node* alignments with *other planets* (in addition to *Mars*);
> *Uranus/Uranus* alignments and other such alignments of *a planet to itself.*

Moving Toward Conclusions

Obviously these findings are not yet sufficiently conclusive, and clearly further astrological reincarnation research needs to be done. However, the findings presented in this book offer a point of departure for further exploration. The research so far has concentrated upon *zodiacal alignments*. The brief look at the Raphael–Novalis *aspect metamorphoses* at the end of chapter 7 seems to indicate that they are not very relevant to astrological reincarnation research. This conclusion is supported by including St. Elisabeth. Looking at the St. Elisabeth–Raphael and also the St. Elisabeth–Novalis aspect metamorphoses shows a remarkable lack of any such metamorphoses, with the exception of the geocentric Venus/Sun conjunction at the birth of St. Elisabeth returning as the geocentric Venus/Sun conjunction at the birth of Raphael. The latter is an *inferior* conjunction, with Venus between the Earth and the Sun, and the former is a *superior* conjunction, with Venus behind the Sun. This could reflect the difference in

their life missions: Raphael (inferior Venus/Sun conjunction) expressing himself more as an artist, and St. Elisabeth (superior Venus/Sun
conjunction) expressing herself more as a healer and saintly nun.

The lack of aspect metamorphoses is striking. Further research
into this is required, however, before coming to any definite conclusions. For example, could it be that conjunctions or oppositions
between two planets A and B could metamorphose and become conjunctions or oppositions between A and C (a third planet)? There is an
example of this in the geocentric chart comparison between the death
of St. Elisabeth and the death of Novalis, as follows:

> † *St. Elisabeth: g-Neptune is opposite g-Mercury*
> † *Novalis: g-Neptune is opposite g-Venus*

Obviously, if aspect metamorphoses occur in this more extended
sense, a more detailed study is required to uncover them—more
detailed than we can undertake in this book

The most important point to note, for now, is that the zodiacal
alignments take place in the *sidereal zodiac*. This is the first (primary) research finding of astrological reincarnation research, and it is
thanks to the Babylonians that the framework of the sidereal zodiac
originally became accurately defined. The second major research finding is that not only the geocentric but also the heliocentric (hermetic)
planetary positions in the sidereal zodiac at conception, birth, and
death are highly significant. The heliocentric chart—and more especially the hermetic chart, which includes the location of the Sun and
Moon, Ascendant and Descendant, Midheaven and the other house
locations, together with the heliocentric positions of the planets—is
an indispensable supplement to the geocentric chart in astrological
reincarnation research.

The foregoing example of three successive incarnations: St.
Elisabeth—Raphael—Novalis offers a striking picture of *metamorphosis* of a great soul (one of the spiritual leaders of humanity: *the
third teacher* of our time) from incarnation to incarnation: from
saintly nun and healer to great artist to great poet and writer. In each
case the earthly life is short, yet powerful and impactful. Behind
these lives stands that of John the Baptist of whom Jesus said: "Truly,

I say to you, among those born of women there has risen no one greater than John the Baptist" (Matthew 11:11). In the light of these words it is appropriate to consider the reincarnations of this great individuality. Here I would like to reiterate that no one is expected to believe that John the Baptist reincarnated as St. Elisabeth of Hungary. For me it is a matter of certainty. Without certainty, however, it can only be a hypothesis, one that may be considered in the light of the karmic biographical background and the astrological reincarnation research presented here.

The volume of research findings is multiplied enormously if we also take the epoch/conception, birth, and death charts of John the Baptist into consideration, as we shall do in the next chapter. However, we are unable to include Elijah, as his dates of birth and death are unknown. This is regrettable as *Elijah—John the Baptist* is the one reincarnation example indicated by Jesus Christ two thousand years ago: *"For all the prophets and the law prophesied until John; and if you are willing to accept it, he is Elijah who is to come"* (Matthew 11:13–14).

10

PHINEHAS AND THE MYSTERY
OF INDIVIDUALITY

THE MYSTERY OF INDIVIDUALITY

IN FOLLOWING THE GOLDEN thread of spiritual intention that lives at the heart of the Elijah stream of incarnations, we come ultimately to the mystery of true individuality. On the path of developing a new star wisdom based on the astrological study of reincarnation, we have an opportunity to discover how the personality and the individuality come to expression in the horoscope. The individuality or higher self can be thought of as a radiant Sun in the realm of spirit. From this Sun a ray shines down and enters into incarnation upon the Earth as the personality. Through the human being's series of reincarnations upon the Earth many rays shine forth from the Sun of the individuality. When a highly evolved individual reincarnates, the ray from the higher self is generally more intense than normal, indicating that more of the higher self comes to expression than is usually the case. The image of a radiant Sun for the higher self is quite real, since the Sun is the spiritual abode of the higher self. Thus the heliocentric (Sun-centered) chart expresses something of the individuality or higher self. Correspondingly, the geocentric (Earth-centered) chart has to do with the level of the earthly personality.

The Spiritual Sun

The Incarnation of Christ in Jesus was the incarnation of the Spiritual Sun, the Self of Selves, into the perfect earthly personality, Jesus. Thus the God-Man, Christ Jesus, was born. Since that time it is possible for the earthly personality—through Christ Jesus—to come into connection with the higher self. In this way the higher self, although essentially bound up with the Sun, is able to participate in the evolution of the Earth.

The hermetic chart—*which is basically the heliocentric chart drawn from a geocentric perspective*—is an expression of the higher self in this Christianized sense. It bears witness to Christ's Incarnation upon the Earth, having come from the Sun. The hermetic chart goes beyond the heliocentric chart, and it is in working with the hermetic chart that it is possible to discover things that are hidden both to the geocentric and the heliocentric chart. Indeed, one meaning of the word "hermetic" is *hidden*.

Baptism in the Jordan—Moon/Mars

As an example of such a hidden factor in the horoscope, let us consider the geocentric, heliocentric, and hermetic horoscopes for the moment of Christ's Incarnation, descending from the Sun, at the Baptism in the River Jordan (see appendix 2, charts 1, 2, 3).

> *Geocentric: Moon* at 16° Leo (16♌16) and *g-Mars*
>   at 4° Virgo (4♍00)—18° apart
> *Heliocentric: h-Mars* at 18° Leo (18♌21) [there is no Moon
>   in a heliocentric chart][119]
> *Hermetic: Moon* (16♌16) conjunct *h-Mars* (18♌21)

Here it is important to observe that neither the geocentric nor the heliocentric chart reveals that there was a relationship between the Moon and Mars at this event. This is evident *solely* in the hermetic chart, and hence the significance of the hermetic chart *in place of* or *in addition to* the heliocentric chart.[120] At the Baptism, descending from the Sun, this hermetic ("hidden") conjunction of the Moon and heliocentric Mars was a *reality* for the God-Man, Christ Jesus, who *bridged* the Sun-centered (heliocentric) and Earth-centered (geocentric) perspectives. Christ incarnated the *living power of the Word*. "We look at the Mars sphere and perceive what is happening there. Everything is *word*. The beings of Mars are *word beings*."[121]

The hermetic conjunction of the Moon and Mars in Leo indicates the *power of the righteously spoken word* (Mars being connected with the sixteen-petalled lotus flower—that of the word—in the region of the larynx).[122]   Mars   16 → larynx

Holy Saturday—A Venus Mystery

Another example indicating the importance of the hermetic chart is revealed by way of considering the *hermetic conjunction* of the *Moon* and *Venus* in Libra at Christ's descent into the underworld on Holy Saturday, following the crucifixion on Good Friday.

> *Geocentric: Midday on Holy Saturday—Full Moon (23♎41)*
> *g-Venus (24♓20) conjunct g-Mercury (20♓59)*
> *Heliocentric: h-Venus (23♎45) square h-Saturn (25♋03)*
> *Hermetic: Moon /h-Venus conjunction (23♎46)*
> *square Saturn (25♋03)*

(The exact Moon/h-Venus conjunction at 23♎46 in the hermetic chart took place ten minutes after noon, local time, on Holy Saturday—see appendix 2.)

Here again, neither the geocentric nor the heliocentric charts reveal the striking relationship between the Moon and Venus at the descent of Christ into the underworld. This is only found in the hermetic chart. The hermetic conjunction of the Moon and Venus at the descent into the underworld reveals the coming together of Earth karma (remembered by the *Moon*) with the activating influence of *Venus*. It is worthwhile to ponder on the conjunction of Christ's descent into the underworld at the time of this hermetic conjunction. Let us consider the following words relating to the Moon and to Venus and then consider them in relation to the descent into the underworld.

> The Moon mystery was one of the deepest secrets in the ancient mysteries, for it is the primal wisdom that the Moon enshrines within itself. What the Moon is able to reflect from the whole universe forms the sum-total of the forces that...sustain the animal element in human beings and are connected with their sexual nature in its physical aspect. So the lower nature of human beings is a product of what radiates from the Moon, while the

highest wisdom once possessed by the Earth lies concealed within the Moon fortress....

Venus is highly responsive to everything that comes from the Earth.... She lovingly reflects whatever comes from the Earth. If with the eyes of soul we are able to glimpse the mysteries of Venus, the whole Earth with its secrets of the life of soul is there before us once again. The truth is that human beings on Earth can do nothing in the secrecy of their souls without it being reflected back again by Venus. Venus gazes deeply into the hearts of human beings, for that is what interests her; that is what she will allow to approach her. Thus the most intimate experiences of earthly life are reflected again from Venus in a mysterious and wonderful way. In the reflection she transforms everything. [123]

The conjunction of the Moon and Venus at the time of Christ's descent into the underworld is a cosmic sign of the motivating impulse underlying this event. It was an impulse permeated with profound cosmic wisdom (Moon), and born out of great love for the Earth and for the mysteries concealed within the depths (Venus), that Christ undertook this journey down toward the heart of Mother Earth.

At the same time there was in the cosmos, viewed from the hermetic perspective, a square of *Saturn* to the *hermetic conjunction of the Moon and Venus.* "Saturn is the preserver of the memory of our universe," and this deed of great love, borne of sacrifice, of Christ's descent into the underworld was witnessed and remembered by the "preserver of the memory of our universe." This heavenly configuration thus serves as a reminder to humanity, to all of us, to uphold our spiritual covenant (karmic intention) to participate—with Christ—in the redemption of Mother Earth, in the creation of the "new Earth" (Revelation 21:1).

Here the God-Man, Christ Jesus, bore the balance and harmony of Venus (in Libra) into the depths of the Earth as a deed of redemption. (Venus is connected with the six-petal lotus flower—that of balance and harmony—in the region of the abdomen). [124] The descent into the underworld was the "sign of Jonah" referred to by Christ: "Then some of the scribes and Pharisees said to him, 'Teacher, we wish to see a sign from you.' But he answered them, 'An evil and adulterous generation seeks for a sign; but no sign shall be given except the sign of the prophet Jonah. For as Jonah was three days and three nights in the belly of the whale, so will the Son of man be three days and three nights in the heart of the Earth" (Matthew 12: 38–40).

One might conjecture that these hermetic (hidden) configurations may well apply to the life of Christ Jesus, who bridged the heliocentric and geocentric levels, but not to ordinary mortals. However, as years of study of the hermetic charts of historic personalities indicate, such hermetic configurations are highly significant. Just to mention one example, Marilyn Monroe (1926–1962), the famous Hollywood film star, was born when there was a hermetic conjunction of the Moon and Venus in Capricorn. As a modern archetype of beauty (Venus), finding acclaim in the magnetic reflective (Moon) world of the cinema, her life as the epitome of the Hollywood woman resembled something of a descent into the underworld, culminating in her suicide through a drug overdose at the age of thirty-six.

Elijah—Transfiguration on Mount Tabor

The counterpart to the Holy Saturday descent into the underworld transpired in the life of Christ Jesus at the transfiguration on Mt. Tabor. Accompanied by Peter, James, and John, Christ went to the top of Mt. Tabor, teaching the three disciples as he went. At the summit, around midnight, the three disciples beheld Him bathed in radiant light, shining like the midnight Sun, between *Moses* and *Elijah*. The hermetic chart for the transfiguration reveals an *opposition* of the *Moon* and *Venus*:

> Transfiguration (midnight): *Moon (23♑15)—opposite—*
> *h-Venus (20♋53)*
> Holy Saturday (midday): *Moon (23♎41)—conjunct—*
> *h-Venus (23♎45)*

The hermetic *opposition* of the Moon and Venus at the transfiguration is complementary to the hermetic *conjunction* of the Moon and Venus in Libra at the descent into the underworld on Holy Saturday.

There is a symbolism inherent in the six-pointed star (representing the six-petal lotus flower corresponding to Venus), wherein two interlocking triangles, one pointing up, reaching up to spiritual heights, and one pointing down, reaching down into the depths, form an interlocking six-pointed star, sometimes referred to as the "Seal of Solomon."[125] This *movement* was demonstrated in archetypal form in the life of Christ by the miracle of the Transfiguration on Mt. Tabor (upward pointing triangle) and the descent into the underworld on Holy Saturday (downward pointing triangle). Christ's life on Earth was a movement in both directions. He maintained a sense of inner peace and harmony wherever he went, this being the spiritual quality of the awakened and developed six-petal lotus flower (Venus chakra). Of course, this was only possible through the perfectly balanced *immaculate* organization of his individuality as Christ Jesus, in the harmonization and stabilization permitted through his divine "I." This was demonstrated as a manifestation of the "I AM THE I AM" proclamation of the living Word, witnessed by the disciples Peter, James, and John on the night of the miracle of the Transfiguration on Mt. Tabor. And it was also demonstrated—to the beings of the underworld—on Holy Saturday at his descent into the depths of the Earth, this being part of the new (Grail) mysteries that we shall return to consider in the last chapter of this book. At both these great (and complementary) events there was a planetary configuration involving the Moon and heliocentric Venus, which are mirrored in the human being in the two lower chakras: the root chakra (Moon) and the six-petal lotus flower in the region of the abdomen (Venus).

The scene of the Transfiguration is directly relevant to our theme of the individuality of *John the Baptist*, who had lived on the Earth several centuries earlier as the prophet *Elijah* (ninth century B.C.E.). For in the brilliant light of the Transfiguration, Christ appeared between Moses and Elijah. It may be recalled that the last painting of Raphael, *The Transfiguration,* was the *signature* piece of Raphael as the reincarnated Elijah. Peter, James, and John, when at midnight on Mt. Tabor they beheld Christ shining radiantly between Moses and Elijah, had a glimpse into a higher realm—that of the great teachers of humanity (known in the East as *bodhisattvas*) who surround the Christ. Moses and Elijah are two *bodhisattva* individualities, who incarnate again and again to bring new impulses into humankind's spiritual evolution for the unfolding of the Christ Impulse.[126]

The Mystery of Moses and Elijah

The mystery of the working together of the Moses and Elijah individualities can be understood more deeply through the "looking glass" of an earlier life of Elijah, indicated by Rudolf Steiner as the Old Testament priest Phinehas, who lived at the time of Moses in the thirteenth century B.C.E. Thus, commenting upon a biblical passage relating to Phinehas from the book of Numbers 25:10–12 (quoted below), and in the course of his discussion of the scene of the transfiguration of Christ, with Moses on one side and Elijah on the other, Rudolf Steiner indicated the following:

> Yahweh announces to Moses that He would bestow on the ancient Hebrew people a special priest to support Him and be united with Him. This priest was Phinehas who, as a son of Eleazar, the son of Aaron, was Aaron's grandson. And the secret doctrine[127] as well as modern esoteric research tell us that there lived in Phinehas' body the same soul later going to dwell in Elijah.... This soul is active in Phinehas. We find it again in Elijah-Naboth, then in John the Baptist; and hence see it make its way across human evolution. This soul is set before our imagination on the one side [of the scene of the transfiguration]; that of Moses himself on the other side. [128]

Phinehas was the grandson of Aaron, the brother of Moses. It was Aaron who served as high priest in charge of the priesthood of the Levis, the tribe who were responsible for the transporting and care of the Tabernacle during the Forty Years of the wandering in the Sinai desert. (Exodus 28:1) Eleazar, the son of Aaron, had succeeded his father as high priest after Aaron's death. Phinehas, Eleazar's son, was still a young man when the people of Israel, while wandering in the Moab desert, were seduced by the Moabite King Balak into the practice of idolatry. Balak's aim was to divert the people of Israel from their spiritual mission by preparing a "banquet before them"—prepared by the Midianites and intended as a sexual orgy. (See also Jude 1:11, II Peter 2:15, and Revelation 2:14 for references to the "teaching of Balaam, who taught Balak to put a stumbling block before the sons of Israel, that they might eat food sacrificed to idols and practice immorality."):

And Israel dwelt in Shittim and the people began to commit for-
nication with the daughters of Moab. Israel invited the people
to the sacrifices of their gods: and the people ate, and bowed
down to their gods. So Israel yoked himself to Baal-Peor: and
the anger of the Lord was kindled against Israel. And the Lord
said to Moses, "Take all the chiefs of the people, and hang
them up before the Lord in the sun, that the fierce anger of the
Lord may turn away from Israel." And Moses said to the judges
of Israel, "Slay every one those men of yours who have yoked
themselves to the Baal-Peor."[129] (Numbers 25:1–6)

In the midst of a devastating plague endemic resulting from a
sexually-transmitted disease found among the Midianites, Phinehas
acted on behalf of the future of Israel:

And, behold one of the children of Israel [Zimri] came and
brought a Midianite woman [Cozbi] to his family in the sight of
Moses, and in the sight of all the congregation of the children
of Israel, who were weeping before the door of the tent of meeting. And when Phinehas, the son of Eleazar, the son of Aaron
the priest, saw it, he rose and left the congregation, and took a
spear in his hand. And he went after the man of Israel into the
tent, and thrust both of them through, the man of Israel, and
the woman through her body. Thus the plague was stayed from
the children of Israel. Nevertheless those that died in the plague
were twenty-four thousand. (Numbers 25:6-9)

And the Lord said to Moses, "Phinehas, the son of Eleazar,
the son of Aaron the priest, has turned my wrath away from the
children of Israel in that he was zealous for my sake among them,
so I did not consume the people of Israel in my jealousy. Therefore
say, Behold, I give to him my covenant of peace: And it shall be
to him, and to his descendants after him, the covenant of an
everlasting priesthood; because he was zealous for his God, and
made atonement for the children of Israel." (Numbers 25:10–13)

The Psalms account of the wilderness journey describes the deed of
Phinehas:

They joined themselves also to Baal-Peor, and ate the sacrifices
of the dead. Thus they provoked him to anger with their doings:

and the plague broke in upon them. Then stood up Phinehas, and executed judgment: and so the plague was staved. And that was counted to him for righteousness to all generations for evermore. (Psalm 106:28–31; cf. also I Maccabees 2:54)

Phinehas served as priest for nineteen years and was the third High Priest of Israel. The final mention of Phinehas is recorded in Judges 19 and 20, involving avenging the sin of the tribe of Benjamin, until the remaining tribes humbled themselves, fasted, and wept before the Ark of the Covenant and the Tabernacle at Bethal. Here Phinehas acted out of righteousness, for the Lord had spoken to him saying, "Go up; for tomorrow I (the Lord) will give them into your hand." (Judges 20:28)

John the Baptist was a descendant of Phinehas, through his father Zechariah, who was struck dumb for his disbelief when an angel announced to him that his wife Elisabeth, who was beyond child bearing age, was to bear him a son (*Luke*, chapter 1). This son, John the Baptist, followed in Phinehas' path of righteousness—his spear of judgment, however, was the living word proclaimed as a judgment toward the king of Israel, Herod Antipas, for all of his evil deeds and for living in *adultery* with Herodias, the wife of Herod's half-brother. John the Baptist was imprisoned for speaking out against Herod, and when a "table was prepared" for a banquet feast for the king's birthday, John the Baptist's severed head was brought out on a platter. Herodias—the granddaughter of Herod the Great and the sister of Herod Agrippa (who was also a brother of Herod Antipas)—had flaunted her daughter Salome as a political move for power. It was Salome's seductive dance before the king that brought the prize of the head of John the Baptist. How similar to the flaunting of Cozbi by Zimri before the sight of Moses and all the congregation of the peoples of Israel that took place at the banquet prepared in the desert by the Midianites?

The Call of Prophecy

Phinehas, as the grandson of Aaron, came from a lineage of prophets. Aaron, according to biblical accounts, had been called by God to be Moses' voice or prophet, for God spoke to Moses and Aaron directly. As the first high priest to be mentioned in the Bible, Aaron is remembered in particular for the blessing that God commanded him to give to the people, "The Lord bless you and keep you and make his face to shine upon you, and be gracious to you, the Lord lift up his countenance on you and give you peace" (Numbers 6:24-26). To be a priest in biblical times meant to be able to bless others and to serve in the Holy Temple.

Why then would Phinehas have been given God's eternal blessing of peace and priesthood after the slaying of Zimri, the Israelite prince, and Cozbi, the Midianite woman, who were engaging in a religious sexual ritual? Perhaps this could be thought of as a spiritual call on the part of Phinehas to set something right in the line of prophecy. One may recall that it was Aaron, the brother of Moses, who earlier had led the Israelites to worship the "golden calf" in the absence of Moses, who had gone up Mt. Sinai to receive the Ten Commandments. Aaron had led the Israelites astray through false prophecy, and it was this that later had its consequences by opening the door to the future temptation in the Moab desert with the religious sexual practices of the Baal cult of the Midianites.

Toward the end of Moses' life, this great leader of the people of Israel who was referred to as the "friend of God" and the "servant of the Lord" once again was threatened with the extinction of his people and their spiritual mission. The Moabite King Balak, fearful of the advance of the Israelites into the Plains of Moab, called upon a false prophet named Balaam to curse the Israelites. A text with Balaam's name and some prophetic curses has actually been excavated in the Iron Age strata at Tell Deir 'Alla in the central Jordan Valley (one of several discoveries of archaeological texts that correspond to biblical accounts).

It was the plains of the Moab desert that provided the setting for the story of Phinehas and the story of sorcery and sexual promiscuity among the Moabites that had once again seduced the Israelites. This "yoking" or joining of Israel to Baal-Peor meant that Israel was genetically compromised, and the leadership of Moses was challenged. Moses himself had married a Midianite woman, and so he was vulnerable to the disappointment of the people through their grief over the loss of loved ones to the plague (24,000) and the executions that, follows upon God's message to Moses to slay all the leaders who had been compromised by relationships with the Midianite women. The incident in the pleasure tent of Zimri, then, was an open challenge to the leadership of Moses who was impotent in the face of it, being responsible for executing 1,000 leaders in Israel and hanging the chief princes in disgrace, for he had been commanded by God to "slay" every other compromised man.

Zimri was a prince of a chief house among the Simeonites and Cozbi was the daughter of a prince of Midian, so that their union was a brazen attempt on their part to compromise the leadership of Moses. The couple had been seduced through the sorcery of Balak, the king of Moab, who through the services of Balaam, the corrupted prophet of the Midianites, had attempted to project spiritual forces against Israel. Zimri paraded Cozbi, "In the sight of Moses and in the sight of all the congregation of the children of Israel." (Numbers 25:6)

"Lord, make his face to shine upon you…and give you peace." This was the blessing of the Lord to the children of Israel given to the people through the prophet Aaron. And it was the blessing of eternal peace that was proclaimed to Phinehas following what must have been judged as a courageous act of righteousness toward the future of his people and their mission, which was to provide a pure vessel for the incarnation of the Jesus individuality. Of Phinehas it is said, "His faith was counted to him as righteousness."[130]

"Behold I give him (Phinehas) my covenant of peace: And it shall be to him, and to his descendants after him, the covenant of an everlasting priesthood; because he is zealous for his God, and made atonement for the children of Israel" (Numbers 25:12–13). Peace in the Jewish tradition is considered to be a sign of the Messianic era and the goal of humanity. This is to be a time when all strife will be reconciled and all creation will be transformed into a symphony of harmony and love. (Isaiah 11:6) We find an echo of this as a remembrance in Novalis's essay "Christendom or Europe." Might we also find strains of this promise in Phinehas' incarnation as Elijah?

The Old Testament Prophet Elijah

There are some very early traditions that suggest that Moses and Phinehas never died, but were taken to heaven, like *Enoch* and *Elijah*. We know through the biblical story of the life of Elijah that he ascended into heaven in a chariot of fire, indicating that he was immortal. Interestingly, there was something living in the collective memory concerning the deed of Phinehas, which was remembered as bearing the mark of immortality. Phinehas, through the swiftness of his *conscious* act of righteousness provided the "shock" necessary to awaken the Israelites—from the path of being consumed by the sexually driven forces of the subconscious senses, representing the past—to a return to an orientation toward the future development of a spiritually oriented human being.

"Elijah does come and he is to restore all things." Thus we have in the Elijah lineage the impulse to set things right in relation to the human being's return to the path of righteousness, calling for a return to the laws of Moses, in a movement forward—toward the development of true individuality through the stabilization of the "I." This represents a progression toward and eventual return to the Spiritual Sun nature of humanity.

When we read the accounts of the life of Elijah, we can recognize this golden thread—of return to a reorientation toward the Future—as the all-encompassing theme in the life of this great and beloved prophet. Remembering that the voice of true prophecy is to bear witness to the living Word, Elijah bore witness to the falsehood and misdeeds of Ahab and Jezebel, who were representative of the fallen nature of humanity, which needs to be shocked into a *recognition* of being driven by subnature into subhuman acts against humanity and thus awakened toward a conscious taking hold of a divinely inspired will development. With the story of King Ahab and his wife Jezebel, we return to the *consequences* of the deed of Aaron, who through false prophecy delivered the Israelites into the worship of the "golden calf," which was a movement back toward the worship of the Bull in the Age of Taurus, and thus signified a leading astray from the path toward the spiritual development of the human being—through overcoming the animal or beastly nature of sexually driven sense reality.

Jezebel like Herodias, and the more unconscious Cozbi before her, could be seen as representing the fallen soul nature, which through greed and lust for power plays a seductive role in influencing the kingly nature to become severed from the *crowning* forces of the head toward the development of conscience, so that the kingly dominion is lost. Ahab, who was the King of Israel in the days of Elijah, coveted what was not rightfully his, the property of another. And although Ahab did seek for the voice of conscience through the prophet Elijah, he was unwilling to hear or believe the voice of truth, and thus was subject to the sorcery and illusory influence of the fallen soul nature, represented as the "apple" offered by the sense driven nature of obsession.

We know from the account of the life of Elijah that he was a witness for truth in speaking out against war, which he believed laid waste to lands and peoples. And that he could bequeath birth and bring the dead back to life, and was attuned to the elements: commanding the waters to part, bringing forth rain, and bringing down the fire from heaven. And in his withdrawal into the desert in his flight from the angry pursuit of King Ahab, whose wife Jezebel wanted Elijah dead, he was completely sustained and supported by nature. These are manifestations of inspiration born of a connection to the living creative substance of the Word.

Ahab lost his kingdom, and Zimri lost his life. One could say that Zimri's behavior was representative of the youth of human development—when one defies the moral teachings of their father and thus risks losing the divine promise of life. In this early developmental stage, the threat of a force of correction comes not from within as a developed voice of conscience, but rather from without. Whereas, moving from the thirteenth century B.C.E., the time of Phinehas and Zimri, to the time of Elijah in the ninth century B.C.E., we see that the further development of the head forces associated with crowning or kingship has given rise to questions of conscience. Ahab sought outside himself for the answer to his question, indicating the beginning of an awakening of conscience, and centuries later Herod *recognized* the voice of conscience in John the Baptist, indicating still further development. Both Ahab and Herod lost their kingdoms, but not without condemning the voice of conscience to death, and, in the case of Herod, even going so far as to sever the developed head forces of John the Baptist, which he feared would bear too great an influence upon the people, and thus weaken his own egotistical dominion.

Thus we see in the life of Elijah a voice of conscience, calling out in the wilderness of humanity's early stages of development. The developing voice for truth, in the story of the unfolding of true human individuality as a vessel for the "I" nature of the Spiritual Sun, is revealed in the life of this great individuality, Elijah, bearing witness that the truth is no longer to be taken up by the sword but rather through the sword of the living Word. It is interesting to note that Moses—who led the Israelites out of Egypt and thus away from the truths of the past into the wilderness of a further developing of consciousness, thus awakening in them a longing for the voice of truth born of true freedom—was not allowed to enter into the promised land, because he had caused water to flow by striking a stone with his rod, rather than by speaking to the stone, as God had commanded (Numbers 20:12-24). Elijah, by contrast, caused the waters of the River Jordan to part and called forth the rain through the power of the living Word.

Mt. Nebo was the final abode of Moses, who could view from the elevated heights of consciousness, the promised land of the future, a future of humanity that was yet to come. It is in the service of this promised future that the mystery of Moses and Elijah, appearing together in the transfiguring light of the risen Christ Jesus, continues on—bearing witness to the living Word of conscience.

The account of the life of Phinehas remains as a *witness* to the work of *vigilance* and *constancy* required of a high priest toward bearing witness to the Holy Name, that it not be profaned in the world. In Joshua 22:13–34 we have an account of the Reubenites and the Gadites being pursued on their journey back to their homelands after battle with the Israelites, because word had arrived that they had succumbed to building a sacrificial alter "at the frontier of the land of Canaan, in the region about the Jordan, on the side that belonged to Israel." Phinehas traveled with ten chiefs, one from each of the tribal families of Israel. Thus came the whole congregation to bear witness:

What is this treachery that you have committed against the God of Israel in turning away this day from following the Lord, by building yourselves an alter this day in rebellion against the Lord? Have we not had enough of the sin at Peor from which even yet we have not cleansed ourselves, and for which there came a plague upon the congregation of the Lord, that you must turn away this day from following the Lord? And if you rebel against the Lord today he will be angry with the whole congregation of Israel tomorrow. (Joshua 22:13–18)

As the story progresses, it is discovered that the Reubenites and the Gadites being fully converted and having repented from their former ways had named the alter "Witness." "For" they said, "it is a witness between us that the Lord is God." (Joshua 22:32)

The Zohar Account of Phinehas

The Zohar, a written manuscript coming from the mystical cabalistic tradition of the Hebrew people, gives a further account of the mystery of the working together of the two witnesses—Phinehas (the previous incarnation of Elijah) and Moses. This account supports our understanding of individuality and the stages of the incarnation of the "I" that can then be followed, thanks to Rudolf Steiner, drawing evermore near with each of the incarnations we are considering in the Elijah stream. Let us first consider the picture that is given us by the Zohar:

Phinehas at that time was not qualified for greatness, but because he was zealous for the name of his Master he rose to the greatest heights *and all was made right in him ... and he was invested with the highest priesthood.*[131]

The words "all was made right in him" refer to the souls of Nadab and Abihu, who united and entered Phinehas at the moment of his great deed.

Here we have our first indication of incorporation not unlike what was to follow with the incorporation of John the Baptist into Lazarus at the raising of Lazarus from the dead, described by Rudolf Steiner,[132] following which Lazarus can be thought of as Lazarus–John. Having passed through death and having been outside of his body for several days, Lazarus had the ability to be outside his physical vessel and thus he was able to spiritually unite with John the disciple at the Last Supper and again at the Crucifixion.[133]

Through the Zohar's account of the life of Phinehas, we are given for the first time an indication of this future mystery:

> Now when Phinehas rose up against the adulterers, when he saw the hosts of the tribe of Simeon gathering around him, his soul fled from him, and then two souls that were flying about naked joined it and they all became one and thus united entered into him, so that he took the place of Nadab and Abihu to become high priest. [134]

This account describes how when a human vessel is made ready through holy zeal and the soul becomes inflamed with a passionate love of God and righteousness, it is possible for the "I" forces coming from the Sun and living in other highly evolved individualities to incorporate their forces in order to ennoble a living personality to carry out a higher covenant to further the destiny of humanity in the path of spiritual evolution.

The narrator speaking through the Zohar credits Elijah with the following account that he (Elijah) was said to have received from the "celestial Academy before the Holy One":

> When Israel left Egypt, God desired to make them on Earth like ministering angels above, and to build for them a holy house that was to be brought down from the heaven of the firmaments, and to plant Israel as a holy shoot after the pattern of the celestial prototype. [135]

So Phinehas, as an individuality acting on behalf of God's desire for Israel to maintain the pattern of the celestial prototype, at this point in our developing story is as one who is a part of fallen humanity, but who stands up in a zealous, righteous act toward a greater covenant in service to the divine plan of creation.

However, to serve further as a high priest for his people, it was necessary that something of a higher nature be passed on to him from one who had previously attained the level of high priest in the service of the people of Israel. This being could clearly be none other than Moses:

> God therefore commenced to speak with Moses, saying: *Phinehas son of Eleazar son of Aaron the priest.* Said Moses: Sovereign of the Universe, what of him? He replied: *Thou didst risk thy life many times to save Israel from destruction, and now he has turned My wrath away from the children of Israel.* Moses thereupon said: What wilt Thou of me? All is in Thy hand. He replied: *Nay, all is in thy hand. Tell him that it (Shekinah) will abide in him.* Moses replied: I am willing with all my heart that it should be with him. God then said: *Declare aloud that thou deliverest it to him with all thy heart.* Hence we read in the text, "Say, Behold I (viz. Moses) give unto him my covenant.... Yet think not that it was taken away from Moses: It was like a light from which another is kindled without loss to itself."
>
> So God persuaded Moses to say, "Behold, I give him my covenant of peace," and until Moses gave it permission to go there it would not go.... So Moses gave the gift to Phinehas on the condition that it should be subsequently restored; and it was through this covenant that Phinehas obtained the high priesthood, and without it he would not have been linked to the grade of the priesthood. [136]

This mystery has something to do with the Ark of the Covenant, which at that time for the people of Israel was carried, honored, and maintained by the priesthood. This covenant was something that would only later be internalized through the activity of conscience in the human being. It is this story that is exemplified through the life of Phinehas:

> As for the tradition that he did not die, the truth is that he did not die like other men, and he outlived all his generation, because he kept this supernal covenant, and when he departed from the world it was with celestial yearning and beauteous attachment. [137]

11

Phinehas – Elijah – John the Baptist

The Mystery of Baptism

It was said that Phinehas "was granted to outlive all those who went from Egypt and won the high priesthood for himself *and his descendants.*"[138]

Why was this grace granted to a human being that had committed murder? According to the Zohar, the answer can be understood through the account of the story of Joshua who was found standing in filthy clothing before the angel of the Lord. And thus, we learn that:

> The evil deeds of man make for him these filthy garments and so God said afterwards to him: *"Behold I have caused thine iniquity to pass from thee and I have clothed thee with rich apparel...."* So Phinehas did not depart from this world until there were prepared for him other garments pleasing to the spirit for the next world.[139]

Here we have the precursor to the mystery of the Elijah stream returning as John the Baptist, for he was preparing the people for the future baptizing of Christ—which was to be the *baptism of grace*, wherein sins are made clean and washed away in the human being who has first repented and undergone a cleansing of consciousness. Through the baptism of grace, the garments are renewed, a regeneration occurs, and the human being becomes a radiant being in the world. Let us follow this story further:

> When a man is inclined to defile himself with sin, he draws down upon himself a spirit from the side of uncleanness that

has a complete sway over him, *but if he makes an effort to purify himself he is helped to do so.*[140]

This is what the baptism by John the Baptist was all about—*a call for penance and purification.* This made possible the *response* from the spiritual world through the baptism by water, which was a drawing down of the divine "charge," the "Matrona" or original matrix of regeneration.

John the Baptist

Now when all the people were baptized, and when Jesus also had been baptized and was praying, the heaven was opened, and the Holy Spirit descended upon him in bodily form, as a dove, and a voice came from heaven, "Thou art my beloved son; today I have begotten thee." (Luke 3:21-23)

What was the mystery that lived so powerfully between the destinies of John the Baptist and Jesus of Nazareth? In Rudolf Steiner's lectures the *Gospel According to Luke,* we are given a breathtaking glimpse behind the veil of spiritual reality—allowing us to further penetrate this mystery:

In the Near East, the state of connection between a higher individuality and a physical body was called "being filled with the Holy Spirit."[141]

The prophets and seers who were guiding Israel were thus said to be filled with the Holy Spirit—that is, guided or inspired by a higher consciousness outside their own "I" capabilities. "Elijah was one of the most important of these seers."[142]

Before the Christian Era—the human body could not embody all the faculties that such a being needed in order to influence humankind.[143]

Elijah was one such "I" being whose I-being could not totally penetrate his physical body.[144]

This remark has to be understood against the background of Elijah's status as a bodhisattva.[145] At this level Elijah was served from without by the overlighting presence of the Buddha, whose on-going service to humanity is toward the internalization of conscience—through the inspiration of compassion. The Buddha himself had served at the bodhisattva level in his previous incarnations—up until his attainment of buddhahood, signifying his last, culminating incarnation on the Earth. However, after his incarnation as the enlightened Buddha, the Buddha was able to influence humankind further from an elevated sphere of existence. Elijah, as one of the twelve bodhisattvas, participated on a spiritual level in the Buddha's attainment of grace. A few centuries after the Buddha's enlightenment, Elijah reincarnated as John the Baptist, and the Buddha, streaming grace from spiritual realms, drew close to the being of John the Baptist, as well as to the Nathan Jesus child—yet "descending only as far as the etheric and astral worlds." [146]

> The normal forces that prompt human incarnation on Earth were inadequate to induce an incarnation such as Elijah's.... External forces influence the developing physical beings to such an extent that the individualities in question incarnate as inspired, ecstatic, or seemingly spirit-driven personalities whose knowledge greatly exceeds the capacity of their ordinary intelligence. Every Old Testament prophet was such a personality. The spirit-driven "I" could not always account for its actions. The spirit lived in the prophet's personality but was sustained from outside....
>
> Personalities like this retreat into solitude from time to time. The portion of the "I" that is used by the personality then withdraws, inducing ecstatic and unconscious states, and the spirit speaks into the person from outside. [147]
>
> The mighty words of Elijah spoken in the ninth century before Christ were actually God's words and his gestures were God's gestures. [148]

As for Elijah's further incarnation as John the Baptist—how did the Buddha's influence affect the destiny of John as the Baptizer?

> John's "I" was further removed from the spiritual world and closer to the Earth than the beings who previously guided

Elijah.... The Buddha's influence actually shaped John's "I" instead of inducing inspiration as had happened to Elijah...so that John's preaching at the Jordan is a revival of Buddha's sermon (at Benares)....[149]

From the mouth of John the Baptist, we hear what the Buddha had to say six hundred years after the end of his final life in a physical body. Such is the unity of all religions.... Refusing to hear the Buddha in John the Baptist is like coming back to a blooming rosebush after having seen it in bud and refusing to believe that the flowers emerged from those buds.[150]

It is important to purify yourself through your own personal forces.... All John's words seem to repeat or elaborate on Buddha's words.[151]

How is it possible that the body of John the Baptist could resonate with such a highly evolved being as the Buddha?

To ensure that from the very beginning, his astral body would be clean and free of all the forces that drag human beings down, John the Baptist was born to an elderly couple in whom desire and passion no longer played a role...the same "I" that was reserved for the Jesus of the Luke Gospel was bestowed on the body of John the Baptist.[152]

This is the mystery that lives in the account of the visitation of the young Nathan Mary (the mother of the Nathan Jesus) to Elizabeth, at that time six months pregnant with John the Baptist. We recall that John was brought into *movement* within his mother's womb during the approach of the Nathan Mary. Elisabeth's child came into movement through receiving the familiar vibration of his own "I" as it's *fulfillment* from the womb of the Nathan Mary, newly pregnant with her divine child, Jesus of Nazareth.[153]

On the night of the Nathan Jesus' birth, it was the Buddha's presence that was experienced by the shepherds who beheld a great light in heaven and heard the words "Peace on Earth to all human beings of good will."

A suitably ennobled human body could be produced only by incorporating the etheric substance taken from Adam, which had remained untouched by all earthly events, into the ether body of this particular Jesus child. This etheric substance was bound up with all the forces that influenced the Earth's evolution before the Fall.[154]

Now we have come further toward an understanding of the interweaving of the destinies of John the Baptist and Jesus of Nazareth—through the sharing of the original "I" forces of the first Adam—John receiving the portion of the "I" of the Adam who lived and walked upon the Earth and Jesus of Nazareth, who received the portion of the divine prototype of humanity, the purified substance that had been held back, never having incarnated upon the Earth until the birth of the Nathan Jesus, whose destiny it was to become the "New Adam."

Through the masterful telling of this hidden background by Rudolf Steiner, we have a glimpse of John the Baptist's work extending well into the future, as a herald for the divine "I" that is the promise and fulfillment of a purified humanity.

The Elijah Individuality—A Sun/Mars Mystery

In our focusing on the task of Elijah, in order to gain an understanding of the mission of this individuality, there is yet another important indication given by Rudolf Steiner:

In Christian esotericism Elijah and Moses are looked upon as those who give the teachings of the two pillars. Elijah was the one who brought the knowledge and message of the one pillar, the pillar of strength; Moses means wisdom or truth, and Elijah means the directing force, that which gives the impulse— it is difficult to express the words in ordinary language. Thus we see these two appear in the spiritualized Earth and, indeed, at the stage of evolution they will then have reached. In the transfiguration of Christian tradition Christ appeared between Moses and Elijah, and this entire procedure appears again at the end of the Earth evolution in such a way that the Sun, the Spiritual Sun of love, the manifestation of the Earth mission of love, appears supported by Sun-Mars and Moon-Mercury, by Elijah and Moses.[155]

This is a reference to the two halves of Earth evolution; the first half *prior to the Mystery of Golgotha*, called Mars, related to the earlier Sun stage of evolution, and the second half *after the Mystery of Golgotha*, called Mercury, connected to the Moon stage of evolution:

> Wisdom and strength have been received as a heritage from previous stages of evolution, from the ancient Moon and the ancient Sun. We shall see that during our Earth evolution this is also expressed by our naming the first half after the representative of the Sun forces, Mars, for we only need note at this point that within our Earth evolution we have in Mars that which implanted iron in the earth; in Mars we see the bringer of strength. And in that which rules the second half of Earth evolution we have the representative of the ancient Moon evolution, Mercury, which embodies in the earth the heritage received from the Moon, wisdom. Thus, the Earth evolution consists of two parts: Mars and Mercury.[156]

We see that *Elijah*, therefore, has a far-reaching mission as the individual who is the "bringer of strength" in the course of Earth evolution, and that he is the human being who embodies the *Sun-Mars* principle of strength.

John the Baptist—Moon/Mars/Mercury

This quality of strength we recognize carried forward in John the Baptist, the reincarnated *Elijah*, who was born at a *conjunction* of Moon, Mars, and Mercury, as described by the seer Anne Catherine Emmerich:

> He was much talked about even in his childhood, for the wonders attendant on his birth were known and a light was often seen around the child. Herod soon laid snares for him, and even before the children's massacre, Elizabeth was obliged to flee with him into the desert...in his sixth year or seventh year, I saw him again led into the desert by his mother...looking back at him she retraced her footsteps, weeping. But the boy appeared wholly unconcerned, and quietly walked deeper into the ravine.[157]

John was never in a school. He was nourished and educated entirely through his surroundings in the desert. He was a child of nature, initiated in the silence of her womb:

> His understanding growing clearer and clearer, learning in an inexplicable way something from everything around him.... I often saw by him radiant figures, angels, with whom he treated fearlessly and confidently, though most reverently. They appeared to be teaching him, directing his attention to different things.[158]

John's *strength* and spiritual substance seemed to be known on a suprasensory level to Herod, whose fury and revenge grew all the more menacing. So much so that John's father Zechariah was eventually captured by Herod's soldiers and was imprisoned on Mt. Zion. There he was murdered, having been tortured and "pierced with a sword, because he would not disclose John's retreat."[159] John's mother, Elizabeth died soon after of grief:

> After which he buried himself still deeper in the desert and thenceforth was altogether alone...he passed from one wilderness to another, I saw him running through the fields by night. He went to that region where long after *I saw John the Evangelist sitting and writing under the high trees*...I saw him in continual and familiar communication with angels, by whom he was instructed.[160]

"Be not uneasy. The child knows well what he is about."[161]

I saw John when full-grown. He was a powerful, earnest man. He was standing in a dry well in the desert, and appeared to be in prayer. A light hovered over him like a cloud, and it seemed to me as if it came from on high, from the water above the Earth. Then a light, shining stream poured over him into the basin below. While gazing at this torrent, I no longer saw John at the edge of the basin; he was in it, the shining water flowing over him, and the basin filled by the sparkling stream....

I saw in the desert in which John dwelt a spiritual church rising up out of the waters that flowed in streams from on high, from Paradise, that floated in clouds, and welled-up in fountains. The church was immeasurably vast.... Far up in the tower, as if on a ladder reaching to heaven, were seen in admirable order the whole host of saints, the entire history of the Promise and the Redemption, and the abodes of the blessed, endless in number. High above all the rest hovered *the Blessed Virgin in a mantle so wide as to cover all.*[162]

This vision of Anne Catherine Emmerich offers a key to the special relationship between *John the Baptist* and the *Virgin Mary,* and through Mary, to the Divine Sophia. In the traditional Russian icon paintings of Sophia, generally the Virgin Mary is depicted to Sophia's right and John the Baptist to Sophia's left—as the two human beings closest to Sophia.

John embodied the strength of the first stage (Mars) and was a herald of the second stage of Earth evolution, the Mercury stage. We see this reflected in the Moon/Mars/Mercury conjunction at his birth. The vivid accounts of John the Baptist, drawn from the clairvoyant recollections of Anne Catherine Emmerich, powerfully support our research into the potency of the steller mysteries. In the lives of the great initiators of humanity, the *strength* of the Sun forces are magnified due to the purity of the earthly human vessel that is destined to carry them. It is clear from these accounts that John's sojourn on the Earth was accompanied by the angelic hierarchy that works from the Moon sphere, and that his Mars forces were disciplined in the womb of silence, the sacred domain of Sophia, and were united with the cosmic understanding weaving in the sphere of Mercury. The purity of his intelligence and his whole life were entirely in the service of the highest ideals of Mars, the moral speaking of the word of truth. The power of his word increased in potency until the day of its ripening had come and his destiny as the baptizer of Jesus and the witness of the Incarnation of the Christ as the living Word was to be fulfilled.

John the Baptist, like Elijah before him, was a *bearer of strength and worked through the power of the Word,* connected with Mars. John's voice was a mighty call for morality, purity, and cleanliness in all matters of life. Like Elijah before him, his was a voice of conscience calling out in the wilderness of humanity's early stages of development.

The following vision received in the silent wilderness of the desert from an earlier time, indicates that *Elijah*, like John the Baptist, also had a spiritual connection to the Virgin Mary. Yet for Elijah in this earlier time, the vision was a revelation of Mary as a pre-incarnatory manifestation:

> I also saw *Elijah* in the desert receive nourishment from the Angels.... I saw all his dealings with Ahab, the sacrifice on Mt. Carmel, the killing of the priests of Baal, Elijah's prayer for rain, and the approach of clouds.... I saw that Elijah called forth through prayer a blessing, from which grew a cloud, and that he guided the cloud, according to inner beholding, so that its downpour was scattered. Otherwise it would have perhaps become a destructive downpour.... He saw in the ascending cloud an image of the *Virgin Mary* and recognized many mysteries connected with her hereditary line and the time of her coming.[163]

Just as John the Baptist leaped in the womb of Elisabeth upon coming into the resonance of Jesus in Mary's womb, so Elijah's heart leaped with joy at the promise he beheld of the hope for the future redemption of humanity through the purity of a cosmic virgin—an early presentiment of the coming of the Rose of the World (described in chapter 9 of my book co-authored with Kevin Dann, *Christ & the Maya Calendar*).

The Five Petals of the Rose

Given Rudolf Steiner's inspiring insight, "Elijah means the directing force...that which gives the impulse,"[164] we can begin to conceive of the mystery of the *directing impulse* of connection streaming through the incarnations that followed the Elijah individuality as the unfolding of a five-petal rose—with Mary–Sophia as the center point.

In each of these incarnations the Sun-like element is apparent. It is as though Sophia is opening her robes—Sophia, the woman who is "clothed with the Sun" (Revelation 12:1). The connection to Sophia that came through Mary provided a source of inner strength for the unfolding of the different dimensions and manifestations in the Elijah stream of incarnations. The close connection with Mary came through in a most remarkable way in the incarnation of the individuality of Raphael, who is famous, above all, for his paintings of the Madonna. Mary was again present in the poetry of Novalis, as referred to earlier, as an inspiration toward the understanding of his beloved Sophia, the source of all true wisdom.

The inspiring force of the Divine Feminine, Sophia, becomes particularly apparent when comparing the Old Testament "prophet incarnations" of Phinehas, Elijah, and John the Baptist with the incarnations in the post-Christian era of St. Elisabeth of Hungary, Raphael, and Novalis, wherein a definite change in the nature and direction of the mission of Elijah emerges. Whereas the incarnations of Phinehas, Elijah, and John the Baptist are recognizably representative as "bearers of strength," there is a progression and change of tone with the later incarnations. St. Elisabeth of Hungary, Raphael, and Novalis were "bearers of healing and beauty." This change of direction has to do, I believe, with the change called forth by the Christ Impulse itself, whereby what was strength in the past will more and more become beauty in the future—not only beauty, but also harmony and healing.

This is because in the far-distant future the ancient Sun (cosmos of strength) will metamorphose into future Venus (beauty, harmony, healing). Against this cosmological background it is possible to understand the new direction taken in the unfolding of the mission of *Elijah* as leading toward this future expression of Venus in the "I" nature of the individuality.

12

THE MYSTERY OF ELIJAH
AND LAZARUS – JOHN

T HERE IS A BIBLICAL account that serves as a telling indicator of this mystery: following the beheading of John the Baptist, King Herod *"was perplexed...because it was said by some that John had been raised from the dead, by some that Elijah had appeared, and by others that one of the ancient prophets had risen. Herod said, 'John I beheaded; but who is this about whom I hear such things?' And he tried to see him."* (Luke 9:7–8, Mark 6:14–16)

Before exploring these connections further, it will help our understanding of this individuality if we bear in mind what took place after the beheading of John the Baptist, whose death took place when he was thirty-one years, seven months old—so that there was still tremendous unspent life energy at the time of his early death. This life energy did not simply dissipate, but became a vehicle for his further activity beyond the threshold of death. He became the inspiring spirit of the disciples of Jesus Christ.

"The spirit of Elijah, which having been liberated from John the Baptist's physical body, affected the disciples as a sort of group soul."[165] John had baptized some of the twelve prior to their having become the disciples of Christ Jesus. These disciples were therefore receptive to the inspiration of *Elijah–John* after he left his physical body, just as they had been while John was in a physical body, having received his baptism.

To further our understanding, let us recall that we find this theme of Elijah serving as an overarching mantle of inspiration flowing also in the earlier account of the lifetime of the prophet Elijah in the ninth century B.C.E., wherein the spirit of *Elijah*, after leaving the body of the Old Testament prophet, rested upon his disciple Elisha.

> Now when the Lord was about to take Elijah up to heaven by a whirlwind, Elijah and Elisha were on their way from Gilgal.... They both were standing by the Jordan. Then Elijah took his mantle, and rolled it up, and struck the water, and the water was parted to the one side and to the other, till the two of them could go over on dry ground. When they had crossed, Elijah said to Elisha, "Ask what I shall do for you, before I am taken from you." And Elisha said, "I pray you, let me inherit a double share of your spirit"... And Elijah went up by a whirlwind into heaven.... Then Elisha took up the mantle of Elijah that had fallen from him, and went back and stood at the bank of the Jordan. Then he took the mantle of Elijah and struck the water, and...the water was parted to the one side and to the other; and Elisha went over. Now when the sons of the prophets who were at Jericho saw him over against them, they said, "The spirit of Elijah rests on Elisha." (II Kings 2:1–15)

It is interesting to note that, according to Anne Catherine Emmerich, the *place in the Jordan where Elijah struck the water with his mantle* was the *place where John the Baptist later baptized Jesus*, at which time the spirit of Christ came down to rest upon Jesus.[166] The passing of the mantle from Elijah to Elisha was an outer sign of the passing of the spirit of Elijah to rest upon Elisha.

Something similar took place approximately one and a half years after the beheading of John the Baptist, when—at the raising of Lazarus from the dead—the spirit of *Elijah–John* came to rest upon Lazarus.

> At the awakening of Lazarus, the spiritual being, John the Baptist, who since his death had been the over-shadowing spirit of the disciples, penetrated from above into Lazarus.... After the awakening of Lazarus, this being is Lazarus–John, the disciple whom the Lord loved.[167]

Thus, for a period of a little more than one and one half years after the beheading of John, the spirit of Elijah–John was active as the group soul of the disciples. This mighty being then came to rest upon Lazarus at his raising from the dead, just as the spirit of Elijah had rested upon Elisha after Elijah's transition to the realm of spirit. Do we find an expression of this astrologically?

Sun/Jupiter Interchange

First, the spirit (or self) of the human being is expressed astrologically by the Sun. Therefore, let us look at the position of the *Sun* at the birth and death of John the Baptist in relation to the chart for the raising of Lazarus. The Sun (12½° Gemini) at the birth of John the Baptist aligned with the *geocentric opposition* between Jupiter (14½° Gemini) and Pluto (13° Sagittarius) at the raising of Lazarus. This is the poignant out-picturing of the "I" forces of John, represented by the Sun at birth, being destined through death to serve in the full flowering of wisdom, represented by the transformative *death forces of Pluto* in opposition to the *wisdom forces of Jupiter* at the raising of Lazarus. Given the above spiritual insights, this can be seen as an act of service pictured in the prevailing planetary configurations.

Birth of John the Baptist – Raising of Lazarus (see appendix 2):

Birth Sun (12♊37)—conjunct—
  g-Jupiter (14♊38) opposite g-Pluto (12♐59)

We see this expressed again in the consideration of the conception chart of John the Baptist, wherein the conjunction of the *Sun* (3° Leo) and *Moon* (4½° Leo) at the *raising of Lazarus* aligns with the position of *Jupiter* (4° Leo) at the *conception of John*:

Conception of John the Baptist– Raising of Lazarus (see appendix 2):

Conception g-Jupiter (4♌18)—conjunct—
  Sun (3♌04) conjunct Moon (4♌39)

This kind of exchange (*Sun/Jupiter* in these two examples) could be interpreted as the spirit of Elijah–John (*Sun*) working upon the creative, active intelligence (*Jupiter*) of Lazarus, and the wisdom of Elijah–John (*Jupiter*) penetrating into the heart (*Sun*) of Lazarus.

"The spiritual being [of] John the Baptist penetrated from above into Lazarus as far as the consciousness soul."[168] This is reflected in the death horoscope of John the Baptist with the *Sun* (14½° Capricorn) aligned opposite both *geocentric Saturn* (17½° Cancer) and *heliocentric Saturn* (16° Cancer) at the raising of Lazarus.

Death of John the Baptist–Raising of Lazarus (see appendix 2):

Death Sun (14♑39)—opposite—g-Saturn (17♋37)
Death Sun (14♑39)—opposite—h-Saturn (15♋56)

This could be interpreted as the spirit of Elijah–John (*Sun*) entering into the consciousness/conscience (*Saturn*) of Lazarus, thereby effecting a "crowning" for Lazarus–John.

Mystery of Initiation—Inspiration

The raising of Lazarus was an archetypal event for humanity, demonstrating the possibility in the future, following Golgotha, for the human being to receive the breath of life from the creative source of the living Word, received as an initiatory breath from the etheric life-body of Christ Jesus. This was the octave of the baptism in the Jordan, wherein a new consciousness was to be born of water, understanding water as the physical agent of cleansing and the faithful bearer of vibratory imprint.

The importance of the desert experience for the Israelites was a movement forward away from the Egyptian mystery tradition where the spiritual aspirant left the physical body during the three-day temple sleep to unite with the spiritual world, which could be born as an imprint thereafter as a source of inspiration throughout life. In contrast, the Israelite's mission under the guidance of Yahweh was to make a step forward, undertaking the work of calling down the spiritual world into the physical body—with the goal of bringing an imprint of the spiritual world into the "I" of the individual.

Moses, in receiving the Ten Commandments, secured for the Israelites a *source* for the necessary cleansing and development of the soul nature, which makes spiritual *communion* possible. The Ten Commandments were given so that new and powerful revelations would flow into the spiritual life of Israel. One can see then just how threatening to the future mission of humanity, which was being carried by the Israelites during the Old Testament days, the temptations were that were presented by the sexual practices of the Baal cult, which would have consumed the human soul's desire to unite with a spiritual orientation leading to the manifestation of true individuality. The Phinehas–Elijah individuality thus acted on behalf of the preservation of the practice of soul purification as a precondition for spiritual union.

John the Baptist stated this clearly as a precondition for his baptism by water, which was to be an act of sanctification for the body, but only if the soul was already purified by right actions. He urged the people to "bear fruits worthy of repentance" and to practice virtue among themselves, one to another. If these preconditions were met, then the water element could become a carrier for the vibratory imprint of divine substance. In the case of Jesus, there was the descent of the Christ into his very being. The baptism of Jesus then came as a fulfillment of the goal of the Israelite's spiritual mission, which was to call down the spiritual imprint from above to unite with the physical body as a guiding source or template toward the development of the "I" of the individual.

The raising of Lazarus was a baptism of another sort. This was a baptism by air, which in the case of Lazarus was a true initiation, a rebirth, made possible through the transformation of consciousness that develops through a conscious undertaking of the work of conscience (Saturn), where the soul begins to work consciously with the spiritual world.

Elijah—Living Source of Inspiration

It is evident that the Elijah–John spiritual being has the capability of working strongly spiritually through another human being or through a whole group as the group soul. Thus, after the death of *St. Elisabeth of Hungary* at the age of twenty-four, so great was her inspiring capacity that she was proclaimed a saint just four years after her death. Raphael's death in Rome was experienced at the Vatican almost like the passing away of a god, and Raphael continued to be an enormous source of inspiration, especially to artists. Again, at the death of Novalis his spirit was experienced as a source of inspiration for the poets and writers of the Romantic Movement, particularly in the German-speaking world.

One important point to bear in mind in relation to the incarnations of this individuality is that we may expect to find incarnations of other disciples from the time of Christ in his karmic circle. This is because of his connection to the disciples, as the inspiring spirit or group soul. Just to consider his incarnation as Raphael, for example, two other outstanding artistic personalities come into view in the circle of those incarnated as Raphael's contemporaries: Michelangelo and Leonardo da Vinci. Could it be that they, also, were reincarnated disciples of Christ Jesus?

Elijah–John—Bearer of Strength

Although there is no secure historical frame of reference for the birth and death dates of John the Baptist, it has been possible, nevertheless, to determine those dates, including the date of his conception. These findings are described in appendix V of *Chronicle of the Living Christ*, which comprises my earlier research into the dating of the life of Christ and the significant events surrounding his incarnation. Also included are the horoscopes of conception, birth, and death for Jesus and Mary. The reader is referred to this work for the way in which the horoscopes for John the Baptist were found.

Making use of the horoscopes presented in *Chronicle of the Living Christ* (see also appendix 2), we see that the geocentric chart at the birth of *John the Baptist* reveals that he was born at a *conjunction of the Moon with Mars and Mercury,* indicating the Mars impulse as a "bearer of strength" as referred to by Rudolf Steiner as the continuing impulse of *Elijah*. At death, when John the Baptist was beheaded at the behest of Herod Antipas, *Mars and the Moon were in opposition to one another.*

Birth of John the Baptist—Death of John the Baptist

Birth conjunction of g-Mercury/g-Mars/Moon—
Death g-Mars opposite Moon
g-Mercury (0♋20)/g-Mars (2♋15)/Moon (3♋15)—
g-Mars (28♓29) opposite Moon (27♍14)

This is an example of an *aspect metamorphosis* from birth (conjunction) to death (opposition) in the same incarnation. Moreover, at John the Baptist's death (†), *Pluto* (12° Sagittarius) was opposite the *Sun* (12½° Gemini) at his birth (✱), i.e., transiting Pluto at death was opposite his natal Sun.

> ✱*John the Baptist Sun (12♊37)—opposite—*
> †*John the Baptist g-Pluto (12♐08)*

The planet Pluto can denote a dark undercurrent in the horoscope, and in the case of John the Baptist, whose death came by way of an actual beheading, the appearance of Pluto opposite the Sun at his birth gives an expressive picture to his death that came by way of black magical forces. John the Baptist, as the messenger and proclaimer of Christ as the "Lamb of God, who bears the sins of the world" (*John* 1:29), was opposed by the workings of black magic through the force of opposition coming from Herod and Herodias. The demand of Salome, at the instigation of her mother Herodias, for the head of John the Baptist on a dish, was a manifestation of black magic.[169]

Let us also note the position of the *Sun at John the Baptist's death*, at 14½° Capricorn. The beheading of John is marked by the fact that the Sun at the Baptist's death was *transiting* the position of geocentric Mars (15° Capricorn) at his *conception*, and was in *opposition* to heliocentric Mars (14° Cancer) at his birth:

> *John the Baptist:* †*Sun (14♑39)—transiting conjunct—*
>   Ø*g-Mars (14♑52)*
> *John the Baptist:* †*Sun (14♑39)—transiting opposite—*
>   ✱*h-Mars (14♋11)*
> [Symbols: epoch/conception (Ø), birth (✱), and death (†)]

The beheading of John brought an end to the powerful activity of the Baptizer and thus an end to the earthly manifestation of spiritual strength that lived in the fiery words (Mars) of the prophet. However, the Sun forces radiating through the "I" of this individuality lived on as a source of inspiration, first for the circle of disciples, and then focused upon Lazarus, following his raising from the dead.

John was above all a prophet with the gift of proclaiming the living Word. This is the highest aspect of the gift of Mars in a horoscope, which is toward the moral speaking out of truth. Biblical accounts describe how the people amassed around John the Baptist and were greatly moved by his words. Thus Herod feared that John's influence over the people might carry them to a revolt, and it was this fear on the part of Herod that brought about the death of the Baptizer.

13

JOHN THE BAPTIST –
ST. ELISABETH – RAPHAEL – NOVALIS

FOLLOWING THE GOLDEN THREAD of connection between the incarnations of Elijah, John the Baptist, St. Elisabeth, Raphael and Novalis, we shall use the horoscopes of Novalis as a template to "look through" in order to unveil how the stars reveal a *remembering* of the unmistakable connection between these individualities.

Novalis's Ascendant-Descendant Axis

Looking at Novalis's charts, first we see the hermetic rule alignment equation:

> *Moon at epoch/conception (14½° Cancer)—conjunct—*
> *Ascendant at birth (14½° Cancer)*

Then we discover this golden thread of connection: the *Moon* at Novalis's *conception* aligned on the *opposite* side of the zodiac to *geocentric Saturn* (15° Capricorn) at Raphael's *death,* and on the *same* side of the zodiac to *heliocentric Jupiter* (13½° Cancer) at the *death* of St. Elisabeth of Hungary and also at the *death* of Novalis. Now including the horoscopes of John the Baptist in our search for alignments, it is most striking that the Sun (14½° Capricorn) at the death of John the Baptist also belongs to this golden thread of alignments extending over a whole sequence of incarnations:

Novalis Epoch Moon (14♋19)

> *—opposite—John the Baptist Death Sun (14♑39)*
> *—conjunct—Elizabeth Death h-Jupiter (13♋29)*
> *—opposite—Raphael Death g-Saturn (14♑58)—*
> *conjunct—Novalis Death h-Jupiter (12♋40)*

The Descendant (opposite the Ascendant) is traditionally connected with death, and it is interesting to see that Novalis's *Descendant* recalls death in two earlier incarnations:

> *Novalis Birth Descendant (14♑19)—conjunct—*
> *John the Baptist Sun Death (14♑39)*
> *Novalis Birth Descendant (14♑19)—conjunct—*
> *Raphael g-Saturn Death (14♑58)*

Summarizing this "looking through" the *Ascendant-Descendant axis* at Novalis's birth to the Moon at his epoch/conception, we see that it *recalls* the memory of *death* in the preceding incarnation as *Raphael*, and in the incarnation prior to that as *St. Elisabeth*, and in the still earlier incarnation as *John the Baptist*, and also that it *prefigures* his death in the incarnation as *Novalis*. This is an example of a golden thread running through a series of incarnations. Moreover, it is worth noting that in the hermetic chart for the *death* of Novalis, a hermetic conjunction of the *Moon* (16½° Cancer) and *Jupiter* (12½° Cancer) exists in the region of Novalis's *Birth Ascendant* (14½° Cancer)[170] and, moreover, that this hermetic conjunction is in the region of the *Death Ascendant* (11½° Cancer).

Novalis's Ascendant at birth (14½° Cancer) *recalls* the deaths of Raphael and St. Elisabeth, in the region of Praesepe, the Beehive (12½° Cancer):

> *Novalis Birth Ascendant (14♋19)*
> *—recalls—h-Jupiter (13♋29) at St. Elisabeth's Death*
> *—prefigures—h-Jupiter (12♋40) at Novalis's Death*

The Sun at Novalis's Birth

The *Sun* (21° Aries) at Novalis's birth aligned on the same side of the zodiac (*conjunct*) with the position of the Moon's Ascending Node (20½° Aries) at the *birth* of *Raphael*, and was *opposite* geocentric Saturn (18° Libra) at *Raphael's birth*. Also, the *Sun* (21° Aries) at Novalis's birth aligned *opposite* heliocentric Mars (21½° Libra) at the *birth of St. Elisabeth of Hungary.*

Novalis Sun (20♈53)—conjunct—
   Raphael Ascending Moon Node (20♈36)
Novalis Sun (20♈53)—opposite—
   Raphael g-Saturn (17♎48)
Novalis Sun (20♈53)—opposite—
   Elisabeth of Hungary h-Mars (21♎34)

The Golden Thread—Sun/Moon/Mars Mystery

Remembering that the hermetic chart reveals hidden truths, we find a golden thread of connection streaming through the incarnations of Novalis, St. Elisabeth, and John the Baptist: with the hermetic *conjunction of the Moon and Mars* at the *birth of St. Elisabeth.* (Since Elisabeth's birth time is unknown, it is not possible to say how exact this hermetic conjunction was—nevertheless, it indicates a relationship to the planet Mars similar to what prevailed at the *baptism of Jesus* in the Jordan, when there was a *hermetic conjunction* of the *Moon and Mars*). Symbol conjunction (♂):

Elisabeth Moon (17♎52) ♂ h-Mars (21♎34)—
Baptism Moon (16♌16) ♂ h-Mars (18♌21)
Symbol conjunction (♂)

Remarkably, at the *death* of *St. Elisabeth* the geocentric position of *Mars* (22° Libra) was almost the same as the heliocentric position of *Mars* at her *birth* (21½° Libra), so that the Sun at Novalis's birth aligned opposite both geocentric Mars from her death and heliocentric Mars from her birth.

Novalis Sun (20♈53)—opposite—
   †Elisabeth g-Mars (22♎18)
Novalis Sun (20♈53)—opposite—
   Elisabeth h-Mars (21♎34)

Further, heliocentric *Mercury* at *St. Elisabeth's death* at 19½° Libra was also *opposite* the *Sun* at Novalis's birth. This, in turn, relates to heliocentric *Venus* at the *birth of John the Baptist* (24° Libra). When we "look through" the position of the Sun at Novalis's birth, we discover a golden thread running through the horoscopes of previous incarnations, appearing once again in a metamorphosed way in the "I" (Sun) of Novalis:

*Novalis Sun (20♈53)—opposite—
†Elisabeth ♄-Mercury (19♎30)
*Novalis Sun (20♈53)—opposite—
*John the Baptist ♄-Venus (23♎53)

The Moon at Novalis's Birth

The position of the Moon at birth remembers our soul karma. Thus, it is remarkable to see that the Moon's position at Novalis's birth (15° Aries) aligns with the Moon (17° Aries) at the ascension of Jesus Christ and also with the Sun (15½° Aries) at the resurrection (see appendix 2). It is not surprising to find a strong relationship to the Christ events in Novalis's horoscope, as his mission from incarnation to incarnation is to be a herald of the Christ. [Similarly, the Moon at Rudolf Steiner's birth (25° Leo) aligned exactly with the Moon's position at Pentecost, clearly indicating Rudolf Steiner's relationship with the Christ Impulse.] In terms of his previous incarnation as John the Baptist, the Moon's position at the birth of Novalis aligned with the heliocentric position of Jupiter (16½° Aries) at the death of John the Baptist. Summarizing these alignment equations:

*Novalis Moon (15♈08)—conjunct—
Ascension Moon (16♈44)
*Novalis Moon (15♈08)—conjunct—
Resurrection Sun (15♈40)
*Novalis Moon (15♈08)—conjunct—
†John the Baptist ♄-Jupiter (16♈24)
[*Rudolf Steiner Moon (24♌48)—conjunct—
Pentecost Moon (24♌59)]

These alignments are particularly striking when we consider John the Baptist's death in connection with the overarching *service of Elijah–John* as a source of inspiration for the disciples and in his union with Lazarus, when Lazarus was raised from the dead. In this way John the Baptist—through the disciples and then through Lazarus—was spiritually present to witness the Christ events.

This lunar remembrance of the Moon in Novalis's birth chart also reveals connections with his previous incarnations subsequent to the incarnation as John the Baptist. The Moon (15° Aries) at Novalis's birth was conjunct *geocentric Mercury* (13½° Aries) and opposite *heliocentric Saturn* (16° Libra) at Raphael's birth. Going back still further, the Moon (15° Aries) at Novalis's birth aligned *opposite* its position (18° Libra) at *St. Elisabeth's birth*, bearing in mind that this Moon position is approximate (for midday on the day of her birth), since we do not know the time of her birth. (See the discussion on accuracy [page 94] concerning the Moon's position when it is an average midday position.)

*Novalis Moon (15♈08)—conjunct—
　*Raphael g-Mercury (13♈16)
*Novalis Moon (15♈08)—opposite—
　*Raphael h-Saturn (15♎55)
*Novalis Moon (15♈08)—opposite—
　*Elisabeth Moon (17♎52)

Novalis—Geocentric/Heliocentric Birth Charts

Let us now proceed further with "looking through" the birth chart of Novalis. Having considered the Sun, the Moon, and the Ascendant at Novalis's birth, let us look at the planets in the geocentric chart and also in the heliocentric/hermetic chart. For a quick overview, easy for the eye to take in, rather than the *alignment equations* used hitherto in this book, we shall in the following make use of a *pyramid arrangement* beginning with the planet in Novalis's birth chart *at the top* and going down the pyramid in the sequence of his previous incarnations, giving the planetary positions rounded to the nearest half-a-degree (for the exact positions see appendix 2):

g-Venus: Novalis's Birth *g-Venus* (3½° Gemini):
opposite g-Uranus (2½° Sagittarius) *birth of Raphael*
also *opposite g-Uranus* (2° Sagittarius) at the *death of St. Elisabeth* and in *conjunction with g-Saturn* (4½° Gemini) at the *birth of St. Elisabeth*

This Venus connection is an out-picturing of the New Testament sequence of incarnations of John the Baptist—St. Elisabeth, Raphael, and Novalis—in a movement forward toward the Future Venus stage of Earth evolution as carried by the *Elijah* stream of inspiration.

The connection that emerges for *g-Mercury* (10½° Taurus) at Novalis's birth is its alignment with the *opposition* between *g-Neptune* (12° Taurus) and *g-Mercury* (9° Scorpio) at the *death of St. Elisabeth.*

> *g-Mercury:* Novalis Birth *g-Mercury* (10½° Taurus)
> Alignment: *g-Neptune* (12° Taurus) *opposite g-Mercury* (9° Scorpio) *death of St. Elisabeth*

> *g-Mars:* Novalis Birth *g-Mars* (13½° Pisces)
> *opposite g-Jupiter* (14½° Virgo) at the *birth of Raphael*
> and *opposite g-Mars* (14° Virgo) at the *birth of St. Elisabeth*

> *g-Jupiter:* Novalis Birth *g-Jupiter* (7½° Aquarius)
> *conjunct g-Mars* (7° Aquarius) at *Raphael's conception*
> *and opposite g-Jupiter* (9° Leo) at *John the Baptist's birth*

> *g-Saturn:* Novalis Birth *g-Saturn* (1½° Leo)
> *conjunct* the *Moon* (0½° Leo) at *St. Elisabeth's death*
> *opposite g-Venus* (2° Aquarius) at *John the Baptist's death*
> *conjunct Descending Moon Node* (0½° Leo) at *John the Baptist's birth*

> *Descending Moon Node:*
> Novalis Birth *Descending Moon Node* (7½° Aries)
> *conjunct* the *Sun* (8° Aries) at the *death of Raphael*
> *conjunct g-Jupiter* (5° Aries) at the *death of John the Baptist*

Novalis: The Outer Planets

Coming now to the outer planets—Uranus—Neptune—Pluto:

> *g-Uranus:* Novalis Birth *g-Uranus* (28° Aries)
> *conjunct g-Uranus* (27½° Aries) at *Raphael's death*
> *conjunct Descending Moon Node* (28° Aries)
> at the *death of St. Elisabeth*

g-Neptune:
Novalis Birth g-Neptune (22° Leo)
opposite g-Venus (23° Aquarius) at Raphael's death.
conjunct g-Jupiter (20° Leo) at Raphael's conception

g-Pluto:
Novalis Birth g-Pluto (29½° Sagittarius)
conjunct g-Venus (27½° Sagittarius) at the death of
    St. Elisabeth
opposite g-Mercury (0½° Cancer) at the birth of
    John the Baptist

Generally speaking, in the case of alignments it is those with the horoscopes of conception, birth, and death from the immediately preceding incarnation that are in the foreground. These, in turn, may go back to alignments with horoscopes from earlier incarnations, yielding a "golden thread" running through successive incarnations in a sequence. This may be continuous throughout a whole series of incarnations, or it may be broken for one or two, or more, incarnations, only to be picked up again later.

Most striking, of course, are the alignments of the same planet—for example, Uranus at Novalis's birth with its position at Raphael's death, and also Uranus at Raphael's birth with its position at St. Elisabeth's death. On the whole, however, it seems that planets interchange with one another, from which we may conclude that it is the astronomical *zodiacal location* of planets against the background of the fixed stars that is of primary importance in discovering the alignment.[171] From this basic finding, it is evident that in the case of astrological reincarnation research it is the revealing of *alignments in the sidereal zodiac* which is the primary factor to be taken into account in the comparison of horoscopes between incarnations.

Continuing research will undoubtedly bring further clarification. Perhaps further research will show with regard to alignments that generally only *certain planets* substitute for other planets. An example of this is provided by the *second "law" of astrological reincarnation,* which states that heliocentric Mercury *and/or* heliocentric Venus at birth in one incarnation tend to align with the positions of heliocentric Mercury *and/or* heliocentric Venus at death in the preceding incarnation. Evidently in this type of alignment heliocentric Mercury can substitute for heliocentric Venus and vice versa.

Novalis—Hermetic Birth Horoscope

Having mentioned this example of zodiacal alignments between heliocentric/hermetic charts, let us "look through" the hermetic birth horoscope for heliocentric planetary alignments, following the same procedure as above in the case of "looking through" the geocentric birth horoscope.

Novalis's Hermetic Birth Alignments

The heliocentric/hermetic planetary positions at Novalis's birth are on the left side of the alignment equations.

Novalis Birth hermetic:

h-Mercury (10½° Virgo)—*conjunct*—
   h-Pluto (13½° Virgo) *death of St. Elisabeth*

h-Venus (15½° Leo)—*opposite*—
   h-Mars (15½° Aquarius) *death of Raphael*
h-Venus (15½° Leo)—*conjunct*—
   h-Venus (15° Leo) *conception of John the Baptist*

h-Mars (17° Aquarius)—*conjunct*—
   h-Mars (15½° Aquarius) *death of Raphael*

h-Jupiter (26½° Capricorn)—*conjunct*—
   h-Neptune (26½° Capricorn) *death of John*
h-Jupiter (26½° Capricorn)—*opposite*—
   h-Jupiter (27° Cancer) *conception of John*

h-Saturn (7½° Leo)—*conjunct*—
   h-Mercury (10° Leo) *conception of John*

Regarding the heliocentric alignments of the outer planets:

Novalis Birth hermetic:

h-Uranus (28° Aries)—*conjunct*—
   h-Uranus (28½° Aries) *death of Raphael*

h-Uranus (28° Aries)—*conjunct*—
  Desc. *Moon Node* (28° Aries) *death of St. Elisabeth*
h-Neptune (23½° Leo)—*conjunct*—
  h-Mercury (24° Leo) *birth of St. Elisabeth*
h-Neptune (23½° Leo)—*conjunct*—
  h-Mercury (25½° Leo) *birth of John the Baptist*
h-Pluto (28° Sagittarius)—*opposite*—
  h-Saturn (25° Gemini) *death of John the Baptist*

Summary—Picking up the Thread

From this analysis we can see how the Sun, Moon, Ascendant, and planetary positions (heliocentric and geocentric) are, generally speaking, based on alignments (conjunctions or oppositions) with planetary positions from the charts of previous incarnations. The accuracy of the findings in this research signifies the importance of the planetary locations in the sidereal zodiac. This has to do with *cosmic memory*, since the sidereal zodiac is an outer expression of the so-called "Akasha Chronicle" in which everything is recorded and remembered. Thus, the deeds of one incarnation are remembered by the planetary positions in the sidereal zodiac, and these positions are picked up again in the next (or in a later) incarnation. So the individual's destiny is literally "carried over" by the stars from one incarnation to the next.

Novalis—Conception Horoscopes

In order to be complete, we shall also "look through" Novalis's conception and death horoscopes to see how they relate to the horoscopes of earlier incarnations. The following is not as exhaustive as the foregoing analysis, since only the most striking alignments are singled out. First, we shall "look through" Novalis's epoch/conception chart.

Novalis Epoch—Conception (geocentric and hermetic):

A-Moon Node (20½° Libra)—*conjunct*—
  D-Moon Node (20½° Libra) *birth of Raphael*

g-Venus (6° Cancer)—*conjunct*—
   Sun (5½° Cancer) *birth of St. Elisabeth*
h-Venus (7½° Gemini)—*opposite*—
   h-Venus (7° Sagittarius) *death of Raphael*
h-Venus (7½° Gemini)—*conjunct*—
   h-Mercury (6½° Gemini) *birth of Raphael*
g-Jupiter (28° Sagittarius)—*opposite*—
   g-Jupiter (26° Gemini) *birth of St. Elisabeth*
Sun (25½° Cancer)—*conjunct*—
   g-Pluto (23½° Cancer) *birth of St. Elisabeth*

In the *hermetic* chart comparison between Novalis's epoch/conception and St. Elisabeth's birth there are several close alignments:

Novalis's conception–St. Elisabeth's birth:

Sun (25½° Cancer)—*conjunct*—*h-Pluto* (24° Cancer)
h-Jupiter (3° Capricorn)—*opposite*—*Sun* (5½° Cancer)
h-Mars (19° Virgo)—*opposite*—*h-Neptune* (18° Pisces)
Ascending Moon Node (21° Libra)—*conjunct*—
   h-Mars (21½° Libra)

Let us now compare Novalis's epoch/conception with St. Elisabeth's death:

Novalis's conception – St. Elisabeth's death:

Sun (25½° Cancer)—*conjunct*—*g-Jupiter* (24° Cancer)
g-Jupiter (28° Sagittarius)—*conjunct*—
   g-Venus (28° Sagittarius)
h-Mercury (0° Virgo)—*opposite*—
   h-Venus (28° Aquarius)
The last alignment recalls the second "law."

Now let us compare Novalis's conception with the conception of John the Baptist:

Novalis's conception – John the Baptist's conception:

g-Mercury (4½° Leo)—*conjunct*—*g-Jupiter* (4½° Leo)

In the hermetic chart comparison between the two conceptions [♂ denotes conjunction]:

Novalis's conception – John the Baptist's conception:

Sun (25½° Cancer) ♂ *ḥ-Saturn (28° Cancer)—conjunct—*
ḥ-Jupiter (27° Cancer)
ḥ-Mars (19° Virgo)—conjunct—Sun (17° Virgo) ♂ *Moon*
(16½° Virgo)

Novalis's conception – John the Baptist's birth

Sun (25½° Cancer) ♂ *g-Saturn (27½° Cancer)—conjunct—*
g-Venus (28° Cancer)

Novalis's conception – John the Baptist's death

Sun (25½° Cancer) ♂ *g-Saturn (27½° Cancer)—opposite—*
g-Neptune (26° Capricorn)

Novalis's Heliocentric – Hermetic Death Chart

Finally, "looking through" Novalis's hermetic horoscope of death, applying a sequential arrangement according to the sequence of incarnations going back in time:

Moon: Novalis Death Moon (16½° Cancer):

> *opposite g-Saturn (15° Capricorn) death of Raphael*
> *conjunct g-Venus (17½° Cancer) conception of Raphael*
> *opposite ḥ-Mars (16½° Capricorn) conception of Raphael*
> *conjunct g-Mercury (18° Cancer) birth of St. Elisabeth*
> *conjunct ḥ-Venus (18½° Cancer) birth of St. Elisabeth*
> *opposite Sun (14½° Capricorn) death of John the Baptist*
> *opposite g-Mars (15° Capricorn) conception of John the Baptist*

Since conception is the moment of the commencement of formation of the etheric (life) body, the Sun at conception (Ø) indicates the seed of life force; the transit of g-Saturn to this seed of life force, at death (†), can be seen as a sign of the withdrawal of life force. Not surprisingly, therefore, the *transit* of *Saturn* to the *Sun's* epoch (Ø) position can sometimes be an indicator of death (†), as it was for the young Novalis:

Ø Novalis Sun (25½° Cancer)—conjunct—
†Novalis g-Saturn (25½° Cancer)

Concerning Jupiter in Novalis's heliocentric/hermetic death chart [♂ denotes conjunction]:

> Novalis Death g-*Jupiter* (2½° Cancer) ♂ *h-Mars* (3° Cancer)
> *conjunct g-Mercury* (4° Cancer) *conception of Raphael*
> *conjunct* g-Mercury (0½° Cancer) ♂ g-Mars (2° Cancer) ♂
>   Moon (3° Cancer) *birth of John the Baptist*

Concerning Mercury in Novalis's heliocentric/hermetic death chart:

> Novalis Death h-*Mercury* (27½° Leo)
> *conjunct h-Jupiter* (27½° Leo) *conception of Raphael*
> *conjunct h-Mercury* (24° Leo) *birth of St. Elisabeth*
> *opposite h-Venus* (26° Aquarius) *death of John the Baptist*
> *conjunct h-Mercury* (25½° Leo) *birth of John the Baptist*

The recurrence of heliocentric Mercury at this zodiacal location (24°–27½° Leo) is very striking: birth of John the Baptist (25½° Leo), birth of St. Elisabeth (24° Leo), and death of Novalis (27½° Leo). *This example clearly demonstrates the great importance of the heliocentric/hermetic chart in relation to the unfolding of the individual's destiny from incarnation to incarnation.*

Saturn marks a "golden thread" in relation to Novalis's hermetic death chart [♂ denotes conjunction]:

> Novalis Death h-*Saturn* (0° Leo) ♂ *h-Venus* (2½° Leo)
> *opposite h-Mercury* (1½° Aquarius) *death of Raphael*
> *conjunct h-Venus* (0½° Leo) *conception of Raphael*
> *conjunct Moon* (0½° Leo) *death of St. Elisabeth*
> *conjunct Descending Moon Node* (0½° Leo) *birth of John the*
>   *Baptist*

These examples should suffice to demonstrate that both sets of charts—the geocentric charts of conception, birth, and death relating to the level of the earthly personality, and the heliocentric/hermetic charts of conception, birth, and death relating more to the level of the solar individuality—are important. Zodiacal alignments stretch from incarnation to incarnation in a "golden thread," which sometimes breaks off for one, two, or more incarnations, only to be picked up again in a later incarnation.

Conclusion

By considering the series of incarnations of the *Elijah* individuality—John the Baptist, St. Elisabeth of Hungary, Raphael, Novalis—definite patterns of zodiacal alignments emerge. The sidereal zodiac, originally defined by the Babylonians, proves itself to be indispensable, as it accurately reveals these alignments between planets from one incarnation to later incarnations. Given the extraordinary number of alignments in the sidereal zodiac of planetary positions between the horoscopes of various incarnations—alignments that would not show up if the tropical zodiac were to be used—it is clearly evident that the sidereal zodiac is the authentic astrological zodiac.

The method of chart comparison—geocentric and hermetic, looking for zodiacal alignments from incarnation to incarnation—can be likened to using a powerful telescope. However, whereas an astronomical telescope is directed to the stars in the heavens, the method of chart comparison is directed to the stars' revelations concerning the incarnations of human individualities upon the Earth.

14

THE COSMIC JOURNEY OF THE SOUL

The Weaving of Karmic Destiny

According to Novalis: When a spirit dies it becomes a human being. When the human being dies it becomes spirit. [172]

Let us reflect upon the significance—for a true understanding of astrology—of the soul's journey in the life after death.

Why does a feeling of grandeur, of reverent awe, come over us when we look up into the starry heavens? It is because without our knowing it the feeling of our soul's home awakens in us. The feeling awakens: Before you came down to Earth to a new incarnation you yourself were in those stars, and out of the stars have come the highest forces that are within you. Your moral law was imparted to you when you were dwelling in this world of stars. When you practice self-knowledge you can behold what the starry heaven bestowed upon you between death and a new birth—the best and finest powers of your soul. What we behold in the starry heavens is the moral law that is given us from the spiritual worlds, for between death and a new birth we live in these starry heavens. A human being who longs to discover the source of the highest qualities they possess should contemplate the starry heavens with feelings such as these. To one who has no desire to ask anything, but lives life in a state of dull apathy—the stars will tell nothing. But if one asks oneself, "How does there enter into me that which is never connected with my bodily senses?" then when one raises one's eyes to the starry heaven, one will be filled with the feeling of reverence and will know that this is the memory of the human being's eternal home. Between death and rebirth we actually live in the starry heavens...That is the mystery of the human being's nature between death and a new birth. After one has gone through the gate of death one expands ever more from the small space of the Earth to the realms of Moon, Mercury, Venus, Sun, Mars, Jupiter, and Saturn. We have then grown into cosmic space, like giant spheres. After we as souls have received the forces of the universe, of the stars, we contract again and carry the forces of the starry world within us. This explains out of spiritual science how in the concentrated brain structure an imprint of the total starry heavens may be found. In fact, our brain does contain an important secret.

We have yet another mystery. As a human being one has gathered oneself together, incarnated in a physical body to which one comes by way of one's parents. After having journeyed far during our expansion in cosmic space, we have recorded our own particular characteristics there. As we gaze from the Earth upward to the heavens, there are not only stars but also our characteristics from previous incarnations. If, for instance, you were ambitious in previous earthly lives, then this ambition is recorded in the starry world. It is recorded in the Akasha Chronicle, and when you are here on the Earth at a particular spot, this ambition comes to you with the corresponding planet in a certain position and makes its influence felt.[173]

That accounts for the fact that astrologers do not merely consider the stars and their motions but will tell you that here is your vanity, there is your ambition, your moral failings, your indolence. Something you have inscribed into the stars is now working out of the starry worlds onto the Earth and determines your destiny. What lives in our souls is recorded in the vastness of space and it works back from space during our life on Earth as we journey here between birth and death. If we truly understand them, these matters touch us closely, and they enable us to explain many things.[174]

The journey between birth and death actually reflects the journey of the soul ascending through the planetary spheres after death. On this path of ascent the soul's karma is inscribed in each sphere, the negative karma in the Moon sphere, the karma of illnesses in the Mercury sphere, the karma of love relationships in the Venus sphere, and so on. Upon returning back down through the planetary spheres, the soul's karmic inscriptions are woven into the astral body on the path of the soul's return into earthly incarnation. The actual working out of how this karma is to unfold as destiny in the new life upon Earth takes place in the weaving of the web of destiny into the etheric body, which is formed between conception (epoch) and birth, as described in detail in *Hermetic Astrology*, vol. 2.

Here it suffices to say that the weaving of the web of destiny into the etheric body between conception and birth mirrors the weaving of the karmic inscriptions into the astral body on the soul's path of descent from the cosmos toward the Earth. The last station on this path of descent is the Moon sphere, and it is here, when the soul is in the Moon sphere, that conception takes place and the etheric body is then woven out of the cosmic ether during the following months until birth takes place. The etheric body, and the web of destiny woven into it, begins to unfold through a rhythm of seven-year periods, starting at birth. This unfolding, in turn, exactly mirrors the soul's path of ascent following death in the preceding incarnation. The first seven years, the Moon period, mirrors the period spent in the Moon sphere in the life after death. Something of our negative karma inscribed in the Moon sphere returns to us as illnesses we undergo during infancy. Likewise, the Mercury karma inscribed in the Mercury sphere plays into the next seven-year period, the Mercury period between the ages of seven and fourteen. Similarly, our inscriptions in the Venus sphere are mirrored—at least in part—in the Venus period of life between fourteen and twenty-one. This continues during the Sun period (21–42), the Mars period (42–49), the Jupiter period (49–56), and the Saturn period (56–63). What we live through after the age of sixty-three reflects what we lived through in the realm of the zodiac, beyond the planetary spheres in the life between death and rebirth. Having depicted the relevance of the journey of the soul in the life after death to an understanding of astrology, let us now continue with a description of this journey.

The Journey of the Soul After Death

As Novalis wrote: "Death is a birth into the realm of spirit."[175]
Viewed in this light, death is something to look forward to, something
to prepare for. It signifies birth into a higher, spiritual realm and the
beginning of a journey toward one's heavenly home.

This journey is described in wonderful, poetic lines by Dante in his
great work *The Divine Comedy*. It is a "journey of the soul" expanding away from the Earth through the realm of the planets in the company of higher spiritual beings (the spiritual or celestial hierarchies).

Rudolf Steiner describes this journey in relation to the soul of
Raphael, following the artist's death in 1520, and prior to his incarnation as Novalis in 1772:

> When we look back into olden times, we see rise up before us
> within the traditions of Judaism the prophetic figure of Elijah.
> We know what significance the prophet Elijah had for the peo-
> ple of the Old Testament, and therewith for all humankind;
> we know how he set before them the goal and destiny of their
> existence. And we have shown how in the course of time the
> being who was present in Elijah appeared again at the very most
> important moment of human evolution, appeared again so that
> Christ Jesus Himself could give him the Initiation he was to
> receive for the evolution of humankind. For the being of Elijah
> appeared again in Lazarus–John.[176]
>
> And further we saw that this being appears once again in that
> world painter who let his artistic power unfold in marvelous
> depths of tenderness, as it moved hovering over the Mystery of
> Golgotha. And we saw how the deeply Christian impulse that
> lived in Raphael, as it were, impelled into color and form the very
> nature and being of Christianity itself. We saw how this impulse
> rose again in the poet Novalis. In the poet Novalis stands revealed
> in wondrously beautiful words what Raphael had placed before
> humankind in colors and forms of rarest loveliness. We see, thus
> following one another in time, beings who are brought together
> into a unity when reincarnation is understood.
>
> We know (for I have often spoken with you of these things)
> how, when the human soul has gone through the gate of death,

it enters the world of the stars. What we are accustomed to call "stars" in the external, physical sense are no more than the outer sign and symbol of spiritual worlds that look down upon us and take their share and part in all the deeds of the evolution of humankind.

We know that the human being passes through the Moon sphere and through the spheres of Mercury and Venus, through the spheres of the Sun and of Mars, and of Jupiter and Saturn. And we know that when, together with the beings of these spheres and together too with other human souls who have also departed from the life on Earth, one has elaborated one's karma, one then turns back again to earthly existence.

Bearing this in mind, let us look for a moment at Raphael and see how he passes through the gate of death, and how he enters the realm of the starry worlds, the realm of spiritual evolution, taking with him the power of his art, which already on Earth shone with the bright light of the stars. We behold, my dear friends, how Raphael enters the Moon sphere, and we see how he comes into association here with the spirits who live in the Moon sphere and who are the spiritual Individualities of the great original leaders of humankind, with whose wisdom Raphael, as Elijah, had been deeply inspired. He meets these Moon Beings, and he meets too all the souls with whom he has lived in earlier stages of Earth-evolution. We see how he unites himself spiritually with the spiritual origin of the Earth, with that World of Being that first made it possible for human beings to exist, and for the earthly to be impregnated with the Divine. We behold Raphael, as it were, completely "at home," united with those whom he had most loved to be with in the Elijah existence, inasmuch as it was they who at the beginning of Earth existence set the goal for the life of this Earth.

Then we behold him wander through the Mercury sphere where, in association with the great cosmic healers, he transforms for his spirituality the power that had been his to create what is so infinitely whole and healthy in color and line. All that he had painted, whether on canvas or as a fresco on the wall, for the help and comfort as well as for the unending inspiration of such as can understand—all his work that was so radiant with

light, showed itself now to him in the great cosmic connection
in which it is able to stand when it passes through the beings of
the Mercury sphere.

And thus was he, who on Earth had unfolded so great a love
for art, whose soul had been aflame with love for color and for
line, transplanted now into the sphere of Venus, which in turn
lovingly bore him across to the Sun, to the Sun existence that
lived in all his incarnations so far as they are yet known to us.
For it was from the *Sun* that he, as the prophet Elijah, brought
to humankind through the medium of his own people the truths
that belong to the goals of existence.

We see how in the Sun sphere he is able to live through over
again in a deep and intimate sense—in another way now than
when he was on Earth as a companion of Christ Jesus—he is
able to live over again what he underwent when, through the
Initiation of Christ Jesus...Lazarus became John.

And all that he has painted in shining light for the followers
of Christ Jesus—he now beholds all this pouring its rays into
the cosmic transformation of the human heart.

And we see further how what he thus had at the foundation
of his life penetrates, wisdom-filled, the sphere of Jupiter. In this
sphere he is able in wisdom to enter into a relation of understanding with such spirits as Goethe—the spirit, that is, that
afterwards became Goethe.[177]

The ascent of Raphael's soul in the life after death through the
planetary spheres is described here briefly, without entering into the
nature of his experiences in the Venus sphere, and without mention-
ing the Mars or Saturn spheres. Then as Dante describes, after the
passage through the planetary spheres: Moon, Mercury, Venus, Sun,
Mars, Jupiter, and Saturn, the human being enters into the realm of
the fixed stars (zodiac).

Kamaloca

This entire ascending journey of the soul is "backward looking,"
where the focus of attention is on the preceding incarnation, at least,
initially. The period in the first cosmic realm, the Moon sphere, is
known as "kamaloca" in the East and as "purgatory" in the West.

Prior to the onset of kamaloca, there is a period of transition from the Earth to the Moon sphere, during which the soul experiences its entire life in the form of a tableau of pictures. It is this experience of a "flashback" of memories that so many people who have had near-death experiences describe. This powerful experience, which begins at the moment of death, generally lasts for about three days. During this time, through beholding the tableau of all the images of what one has lived through in life, a strengthening of one's sense of personality takes place. The tableau of memories becomes imprinted into the soul, so that nothing of one's life experience is lost. This is the significance of the words "eternal memory" sung in the funeral rite of the Orthodox Church. Gradually the tableau expands and is imprinted into the cosmic ether (Akasha Chronicle), fading from view about three days after death. Then kamaloca (purgatory) begins, which consists of a re-living of one's life in reverse direction (first the last day of one's life, then the last-but-one day, and so on, continuing until the day of one's birth). Purgatory is not simply a reliving of experiences; it comprises a moral evaluation of everything that one has thought, said, or done. Kama means "desire," and kamaloca is the place of purification from desires. It is where our negative karma is inscribed, cosmically speaking, in the Moon sphere, which is the abode of the Angels in Dante's description in *The Divine Comedy,* and this is confirmed by Rudolf Steiner.[178] The light of our Guardian Angel shines into us here as fully awakened conscience, with which we behold our deeds (and misdeeds).

Each night, between falling asleep and awakening, we relive the experience of the preceding day in the form of a moral evaluation, together with our Guardian Angel working from afar. In the life after death we relive this experience of the night, now with the Guardian Angel near to us, so that our conscience becomes fully awakened. Kamaloca is a re-living of our "night experiences" and thus lasts as long as we were asleep during earthly life—usually for approximately one-third of the length of life.

The night experience is a reliving, as a moral evaluation, of the experience of the preceding day. If we hurt someone during the day, either through thought, word, or deed, at night we relive this from the standpoint of the person we harmed. Thereby we morally evaluate our actions, our words, and our thoughts from the preceding day. This takes place in the reverse order of our day experience. Thus, if we had been very nice to someone immediately after awaking, but had wounded them through our words just prior to going to sleep, the night experience of re-living this in reverse would start with our experience of being wounded and would conclude with being bathed in gratitude, as the experience of being the *recipient* of what we put forth from ourselves. We have little or no recollection of this night experience of moral evaluation of the day as long as our conscience is not fully awakened.

After death, however, through the nearness of our Guardian Angel, the full light of conscience shines into the re-living of our accumulated night experiences. The effect of experiencing in the full light of conscience the consequences of the negative karma of our thoughts, words, and deeds is that we resolve to make good the harm we have done. The resolution made together with our Guardian Angel to put right our misdeeds forms part of our karmic intentions for the next incarnation, the part we can summarize with the words: "Making good our negative karma." Then we lay aside our negative karma—our "karmic cross"—in the Moon sphere, when we have completed the re-living of our accumulated night experiences comprising kamaloca. This karmic cross remains in the Moon sphere until we return to the Moon sphere on the path of descent to a new earthly incarnation, when it is taken up again. Each one of us has to "carry our cross," the cross of our negative karma, when we return to the Moon sphere on our path of reincarnation. The experience of kamaloca in the Moon sphere in the first period of life after death differs greatly from person to person. It is the time when the "veil" is lifted and the hidden side of our earthly life and existence is revealed.

The Raising of the Veil

There are different levels to the raising of the veil from our earthly life. The first level is what we experience in the Moon sphere. The Moon is like a mirror in the cosmos, reflecting everything taking place. Entering into the Moon sphere can be likened to "stepping through" this cosmic mirror, rather like "Alice through the looking glass." We enter a world in which we experience ourselves mirrored back as others experienced us. In earthly life we feel ourselves to be at the center of our world, and it is difficult for us to get outside of ourselves to know how others experience us. This aspect of our existence is veiled from us. The moment we step through the cosmic mirror, the veil is lifted. Instead of experiencing ourselves centrally, we have a "peripheral experience." Everything that proceeded from us in thought, word, or deed is mirrored back so we experience ourselves on the receiving end. This is the significance of the deep wisdom underlying the maxim: "Do as you would be done by." Generally this entire experience is accompanied by a deep feeling of shame. However much we may have thought ourselves to be wonderful or beautiful during life, when the mask is removed, we experience ourselves as we truly are, and we feel ashamed of ourselves. This level of the raising of the veil—in the Moon sphere, during kamaloca—is where we become aware of the effects of our wrongdoings in thought, word, and deed.

Upon completion of kamaloca in the Moon sphere, we pass into the Mercury sphere. Here a new level of the lifting of the veil is experienced, whereby we become aware of the karmic consequences of our illnesses from earthly life. When we are ill, this influences our ability to function in the world and impacts the people around us. This has nothing to do with wrongdoings, at least, not in a direct sense. On a deeper level our illnesses are generally brought about either by our own foolishness, lack of foresight, wrong habits and attitudes, and so on, or they are the consequence of wrongdoings wrought in earlier incarnations. An exception to these two possibilities is when someone as a *sacrifice* takes on some illness or disability, the highest archetype of this being Christ Jesus, who took the "sins of the world" upon himself. Also, the sins of parents may be visited upon their children, grandchildren, and later descendants, in the form of illness or disability.

In the Mercury sphere we experience our illnesses and disabilities from the "other side." During earthly life, when we are ill, we have an experience of being at the center and being cared for by others, and generally we are unaware of the spiritual dimension of our illness. In the Mercury sphere we experience the mirror image of this: from the periphery—how others experienced us during our illnesses. Naturally there arises the impulse to make good and to care for those who cared for us. This new level of experience, the karmic consequences of our illnesses, is under the supervision of the Archangels, who dwell within the Mercury sphere, as described by Dante and also by Rudolf Steiner.[179] Together with the Archangels we make our resolve to make good the karmic consequences of our illnesses. Just as our Moon karma, our negative karma, is inscribed into the Moon sphere as our "karmic cross," so our Mercury karma is inscribed into the Mercury sphere and awaits us there until we return back to the Mercury sphere on our path of descent toward the Earth, when we reincarnate. In the Mercury sphere, through the light cast by the Archangels, the spiritual dimension of our illnesses is revealed to us. Thereby a deeper level of healing takes place. The Archangels are "cosmic healers" who remove our illnesses from us, allowing the spiritual consequences of our illnesses to radiate out and become absorbed by the cosmos in a process of cosmic healing.

Upon completion of this cosmic healing we pass into the Venus sphere and into the company of the beings indwelling this sphere—the Archai, or Time Spirits (Principalities, according to Dante).[180] Here a new level of the lifting of the veil from our earthly life begins, and we are purified of the consequences of our love relationships. The mask veiling our feeling life is removed. This can best be explained in terms of "animus" and "anima" from Jungian psychology. Every man has a female side, his anima, and every woman has her male side, her animus. Generally speaking, love between a man and a woman occurs when the man's anima finds a resonance in relation to the woman and the woman's animus finds a resonance in relation to the man. The woman recognizes in the man the masculine ideals of the male side of her being and is attracted toward this. Similarly, the man perceives in the woman the feminine qualities of the female nature of his own being and is likewise attracted to this. This "explanation" of male-female love is oversimplified, of course, since there are also karmic factors at work when two people feel love for one another.

Venus and Karmic Destiny

Mention was made in chapter 5 of the love of Raphael's life, la Fornarina, the baker's daughter who served as a model for many of his paintings. His love finds expression in his tender paintings of the Madonna, most especially in the monumental "Sistine Madonna" painting that hangs in the Dresden Museum in Germany. The great love of Novalis's life was Sophie von Kühn, whom he met on November 17, 1794, when he was 22½ years old. This meeting changed the course of his life, and gave birth to the poet within him. The unusual nature of his response to meeting the twelve-year-old Sophie can leave no doubt that this was a karmic meeting.

> "In a quarter-hour my mind was made up," Hardenberg [Novalis] insisted to his brother Erasmus. How could such a remark be taken seriously at the time? Sophie was not yet thirteen years old, her womanhood unawakened, a school child who had difficulty with her spelling, naively and uncertainly putting down her words as they were pronounced in the dialect of Saxony. She was ten years younger. . . .
>
> The spontaneity of her sunny, childlike nature and the grace of her presence enthralled Hardenberg [Novalis] from the first moment, but they do not explain the awakening of his soul and the consecration of his spirit that sprang from their meeting and gave focus to his whole life. . . . It is impossible to believe that his soul had ever before been touched by such inexorable faith in the destiny of love such as Sophie's, which overwhelmed his whole being and all his senses. All at once he was aware of it. "What birth denied me, good fortune has given into my hands. In my own family circle I missed what I see gathered in this company, so strange to me. I feel that there are affinities closer than blood."[181]

Here Novalis brings to expression that karmic ties are closer than blood ties. He made this reference to Sophie's stepsister, who was part of "this company, so strange to me." On her thirteenth birthday Sophie and Novalis became secretly engaged, but their engagement was not announced to her family until her fourteenth birthday, after she had just recovered from a serious illness. A few months later the illness erupted again, and two days after her fifteenth birthday Sophie died.

Novalis was shattered by this event, and he resolved to die in her wake, which actually led to his own "higher birth."

> Her death became the rebirth of his spirit...His being had found its focus. The earthly name of the departed bride fused with the symbol of the heavenly Sophia and gave birth to his mystery of love by which Friedrich von Hardenberg became the poet Novalis.[182]

Novalis did not attend the funeral of his young fiancée, and he did not visit her grave until four weeks after her death, on Easter Sunday, April 16, 1797. Two days later he began to keep a journal of his questions, his seeking, and his experiences. His entry into the journal for May 13, when he visited her grave again, is particularly significant:

> In the evening I visited Sophie. I was indescribably happy there. There were flashes of ecstasy. The grave wafted before me as though it were dust. Centuries passed by in a flash. Her presence was tangible—I felt she might appear at any moment.[183]

This "vision at the grave" (initiation would be a better word than vision) has come to be seen as the real birth moment of Novalis the poet. Aspects of this experience recur in the third poem of his cycle *Hymns to the Night*.

Novalis's Birth as a Poet

Let us look, therefore, at the planetary configuration on this day in relation to Novalis's horoscope of birth (chart comparison, appendix 2: Novalis Birth/Gravesite Initiation). Brief mention was made in chapter 4 of one significant element from this chart comparison. Now we shall consider further elements. In the *geocentric* chart comparison we see that on the evening of the vision at the grave there was an opposition of the *Moon* and *Saturn* along the axis where *Venus* was located at Novalis's birth, such that Saturn was in conjunction with the Venus birth location and the Moon was opposite [♂ is the symbol used here for opposition]. In astrological terms this is referred to as a *transit* of

Saturn across the birth position of Venus. Also, the planet *Venus* was transiting the position of *Uranus* at Novalis's birth:

Geocentric Comparison Chart:

Novalis Birth/Gravesite Initiation (Sophie von Kühn):
g-Venus (3½° Gemini)—conjunct—
g-Saturn (4½° Gemini) ☍ Moon (4° Sagittarius)
g-Uranus (28° Aries)—conjunct—g-Venus (27½° Aries)

In the *hermetic* chart heliocentric *Venus* was transiting the *Sun's* location at Novalis's birth. Here Venus emerges as the cosmic muse at the birth hour of the poet Novalis, whereas Sophie was the earthly muse, removed to the spiritual realm and inspiring from spiritual heights (see chart on page 40):

Heliocentric Comparison Chart:

Novalis Birth/Gravesite Initiation (Sophie von Kühn):
h-Venus (21½° Aries)—conjunct—Sun (21° Aries)

> Venus has a great deal to do with poets. Only the poets do not know this, of course, but she has much to do with poets.... If one wants to get to know the cosmic impulses underlying the formation of human speech, then one has to look to this peculiar life and weaving that takes place between Venus and Mars. Thus, if destiny is played out in this way, it is highly significant for the development of a people's language how Venus stands in relation to Mars. For example, if Venus stands in square to Mars, a language becomes inwardly deepened, full of soul.
>
> On the other hand, if Venus and Mars stand in conjunction, and if this then has an influence upon the people concerned, a language becomes soulless, shrill.[184]

These remarks about the influence upon language of the relationship between Venus and Mars are made with respect to an entire people. Astrosophical research into the horoscopes of poets shows that they also apply to an individual, albeit poetic talent seems to thrive in the case of a conjunction, opposition, or square between Venus and Mars. In other words, it is the fact of a relationship (aspect) between

Venus and Mars *as such* that seems to be important for furthering poetic and literary talent.

In the case of Novalis, it is striking that he was born at the time of an exact *opposition* between *Venus* and *Mars* in his hermetic chart. This relates more to the level of the higher self. Thus, it could be said that this aspect between Venus and Mars in his hermetic chart indicates Novalis's latent poetic talent. Yet for this to become born, something of his earthly personality had to die, which it did when after Sophie's death he pledged to "die in her wake." Following Sophie's death, poetry became, for Novalis, his way toward healing:

> Poetry is the great art of constructing transcendental health. The poet, therefore, is the transcendental physician. Poetry...mixes everything for its great purposes—*the exultation of man above himself.*[185]

At Sophie's death heliocentric (hermetic) *Venus* was at 23½° Capricorn opposite the *Sun* (25½° Cancer) at Novalis's conception (see appendix 2 for the comparison chart of Sophie von Kühn's death horoscope to Novalis's epoch/conception chart). Eight weeks later, at the "grave experience" heliocentric *Venus* had progressed through one-quarter of the zodiac, to 21½° Aries, to *transit* the *Sun* (21° Aries) at Novalis's birth. This reveals the initiation experience undergone by Novalis in relation to his higher self (inner Sun) coming into a relationship with the Divine Feminine (Sophia) through his beloved Sophie, indicated in the cosmos by the transit of heliocentric Venus to Novalis's birth Sun. This initiation experience activated the latent poetic faculty of language indicated by the Venus-Mars opposition in the hermetic chart at Novalis's birth.

> *Sophie Death h-Venus* (23½° Capricorn)—*opposite*—
> *Novalis Epoch Sun* (25½° Cancer)
> *Gravesite Initiation h-Venus* (21½° Aries)—*conjunct*—
> *Novalis Birth Sun* (21° Aries)

We should also note that the *Ascending Moon Node* (2½° Gemini) transited the birth position of *geocentric Venus* (3½° Gemini) at the time of Novalis's first visit to Sophie's grave on April 16, 1797, four weeks prior to his initiatory "grave experience." In fact, this transit of the Ascending Moon Node to g-Venus at Novalis's birth was taking place the whole time between Sophie's death on March 19, 1797, and Novalis's first visit to her grave exactly four weeks later:

From Sophie's death to Novalis's first visit to Sophie's grave:
Ascending Moon Node (2½° Gemini)—conjunct—
Novalis Birth g-Venus (3½° Gemini)

Regarding the lunar nodes as "windows to the astral world"— to use an expression derived from Rudolf Steiner's indications[186]— as can be seen from the above alignment equation, these gateways were open and pouring through their inspiration at this time particularly in relation to Venus in Novalis's life, so that the sphere of love and emotions (the Venus realm within and also the realm of poetry) were activated.

Sophie and Novalis's Chart Comparisons

Comparing Sophie's birth chart with that of Novalis, we see that *g-Venus* (11° Pisces) at Sophie's birth was conjunct *g-Mars* (13½° Pisces) at Novalis's birth, which is a clear picture as to how Sophie's Venus nature united with Novalis's creative expression (Mars, the power of speech).

Sophie Birth g-Venus (11° Pisces)—conjunct—
Novalis Birth g-Mars (13½° Pisces)

Further, the *Moon* (13° Aries) at Sophie's birth was close to the *Moon* (15° Aries) at Novalis's birth. (As we do not know Sophie's time of birth, the position of the Moon used here is that for midday on her day of birth, and so the actual Moon position could differ by up to six degrees from 13° Aries.)

Sophie Birth Moon (approx. 13° Aries)—conjunct—
Novalis Birth Moon (15° Aries)

In the comparison of their hermetic birth charts, there is a conjunction of *h-Mercury* (8° Virgo) at Sophie's birth with *h-Mercury* (10½° Virgo) at the birth of Novalis. (Again, the actual position of heliocentric Mercury at Sophie's birth might differ from 8° Virgo by two degrees in either direction.)

> *Sophie Birth ☿-Mercury (8° Virgo)—conjunct—Novalis Birth*
> *☿-Mercury (10½° Virgo)*

This recounting from the life of Novalis and his love, Sophie, is to illustrate the influence of the planet Venus and the karmic memory resulting from prior incarnations. It is in the realm of Venus that, through the activity of the Archai (Time Spirits), light is shed upon the mysteries of love relationships. The veil is lifted, and we see the true, underlying nature of the relationships and what they signify in terms of "accumulated karma" that needs to be set right. This Venus karma is then inscribed into the Venus sphere, waiting to be picked up on the path of descent back into incarnation. The work of the Archai in the Venus sphere consists in unmasking the real nature of romantic love (whether it is "true love" or whether it is a matter of some kind of "projection," or whether it is a mixture of things). To the extent that love is not completely pure, a purifying of the soul takes place in the Venus sphere so that "true love," which is selfless in nature, may prevail in the place of egotistical impulses. Just as in the Mercury sphere, through the Archangels, *cosmic healing* takes place, so in the Venus sphere, through the Time Spirits, we become bathed in *pure, cosmic love* that heals us of any traces left of egotistical aspects interwoven into our relationships from earthly life.

The Sun Sphere—Realm of the Higher Self

After the Venus sphere we enter the Sun sphere of existence, leaving behind our imperfections inscribed in the Moon, Mercury, and Venus spheres. The Sun sphere is a realm of pure goodness; no evil or negativity can enter here. It is here that we become united with our true essence, the higher self, whose home is within the Sun. The higher self is purely moral, and in light of the higher self we are able to gain an overview of the preceding earthly incarnation from a moral perspective, to evaluate the extent to which our incarnation unfolded in accordance with the great ideals of human and cosmic evolution.

On the basis of this experience our Sun karma becomes written into the Sun sphere. The Sun karma comprises essentially the resolve to live more in accord with the higher self and these future ideals. Sun karma is born through becoming aware of the discrepancy between our higher self and the actual life lived on Earth. The inscription of the Sun karma extends through three levels corresponding to the three celestial hierarchies indwelling the Sun: the Exusiai, Dynamis, and Kyriotetes (Powers, Mights, Dominions, in the language of Dante).[187] The *Exusiai* (Hebrew *Elohim*, or *Powers*) are the "powers" of form—having to do with the morality of bringing form into the world (they are also called "Spirits of Form"). The *Dynamis* are beings of movement (they are also called "Spirits of Movement"), and the *Kyriotetes* are beings of wisdom (they are also called "Spirits of Wisdom"). Therefore the Sun period in life embraces three seven-year periods, from twenty-one to forty-two, reflecting the Sun karma inscribed in three levels in the Sun sphere.

Whereas the passage of the soul through the Moon, Mercury, and Venus spheres is a journey through three levels of the soul or astral world, the passage through the three levels of the Sun sphere is a journey of the human being through the spiritual world known as "devachan," or what is known in the Christian tradition as the "kingdom of heaven."

Upper Devachan

When the human being enters and ascends through the Mars, Jupiter, and Saturn spheres, this comprises the passage through the higher spiritual world called "higher devachan."

In the Mars sphere, with the help of the *Thrones*, the spiritual beings indwelling this sphere, memories of earlier earthly incarnations begin to arise. Here begins the "great review" of the sequence of earthly lives, in contrast to the "small review" of the immediately preceding incarnation. Each incarnation is experienced as a "creative deed," and now the task is to grasp the significance of these isolated "deeds" in their totality. The discrepancy between the actual *deed* of our previous incarnation in relation to the ideal underlying the overall unfolding of the sequence of incarnations is perceived with the help of the Thrones. The resolve to make good this discrepancy comprises our Mars karma, which is inscribed in the Mars sphere and mirrored in the Mars period of life between 42 and 49.

In the Jupiter sphere we elaborate our karma together with the *Cherubim*, the beings indwelling this sphere. It is here that we gain an overview of the wisdom underlying the sequence of previous earthly lives. We begin to perceive the *meaning* of the various incarnations in their relationship to one another and to the whole sequence. Here again the discrepancy between the higher wisdom of the totality of incarnations and the actual wisdom (or lack of wisdom) underlying the immediately preceding incarnation, is revealed to us through the light of the Cherubim, and the resolve to make good this discrepancy constitutes our Jupiter karma inscribed into the Jupiter sphere. The influence of the karmic inscription in the Jupiter sphere makes itself felt during the Jupiter period of life between the ages of 49 and 56.

Then, in the Saturn sphere, we work upon our karma together with the *Seraphim*, the spiritual beings indwelling the Saturn sphere. Here there takes place a beholding of the tableau of all earthly incarnations from a *moral* perspective. We realize, with the help of the Seraphim, how much (or little) of our preceding incarnation on Earth was aligned with the higher moral demands of world existence. We perceive the sum total of the past incarnation from a moral perspective against the background of the tableau of the entire sequence of reincarnations. Perception of the lack of alignment with higher morality leads to the resolve to make this good, thus forming our Saturn karma inscribed into the Saturn sphere. This, in turn, influences us during the Saturn period of life, between 56 and 63.

After completing the passage through the ranks of the nine spiritual hierarchies (Angels, Archangels, Archai, Exusiai, Dynamis, Kyriotetes, Thrones, Cherubim, Seraphim), which is at the same time the journey of the soul through the planetary spheres (Moon, Mercury, Venus, Sun, Mars, Jupiter, Saturn), the human being—now as a pure spirit—enters the divine world, the realm of fixed stars (zodiac). Then there takes place the meeting with the Father, in the realm of the Holy Trinity, a meeting that is not usually remembered, except by the great leaders of humanity, such as the one who is called the "Friend of God" (Master Jesus).[188] This moment in time—

approximately the midpoint between incarnations—is referred to as the "midnight-hour of existence."

The Spiritual Hierarchies

Those who recall Rudolf Steiner's statement quoted in chapter 1 that the zodiacal constellations are the outer signs of the spiritual beings of the first hierarchy (Seraphim, Cherubim, Thrones) may find this statement to be contradictory to the foregoing description. How may Rudolf Steiner's indication be understood in the light of the preceding outline of the human being's journey of ascent through the planetary spheres in the life after death?

Briefly, it can be said that the beings of the first hierarchy (Seraphim, Cherubim, and Thrones) indwell the divine world in direct proximity to the Godhead, the Holy Trinity. They are *zodiacal beings* who, however, work down into the upper planetary spheres (Saturn, Jupiter, Mars), where they help human beings with the elaboration of their karma when passing through these spheres. Thus, on the one hand:

> The Seraphim, Cherubim, and Thrones...enjoy "the immediate sight of the Godhead..." We must, therefore, picture the activity of Thrones, Cherubim, and Seraphim, who spiritually represent the zodiac.... In that direction over there are the Thrones, Cherubim, and Seraphim known as the Twins [sign of Gemini], over there, the Lion [sign of Leo], etc. The constellations [sidereal signs] of the zodiac are more than mere signposts but, as a first stage, it is important to realize that, when we refer to the zodiac, we are speaking of spiritual beings [Seraphim, Cherubim, and Thrones].[189]

Here it is clearly stated that the Seraphim, Cherubim, and Thrones are zodiacal beings. On the other hand:

> When a human being enters these spheres, having passed through the Sun existence, the beings of the higher hierarchies begin to work manifestly: first the Thrones in the Mars sphere; then the Cherubim in the Jupiter sphere; and the Seraphim in the Saturn sphere.[190]

In other words, although they are primarily zodiacal beings, the Thrones, Cherubim, and Seraphim work down into the Mars, Jupiter, and Saturn planetary spheres. The activity of the beings of the second hierarchy (Exusiai, Dynamis, Kyriotetes) is *within* the Sun sphere, and therefore the reflection in the life of the human being of the passage through the Sun sphere, during the Sun period (21–42), is three times longer than the other seven-year periods connected with the planets.

> The period from the twenty-first to the forty-second year needs a survey three times more comprehensive than before, because here all the beings of the Sun mysteries are revealed.... In the corona of the Sun there are incandescent gases, but within this empty space there is nothing physical, not even space! It is all purely spiritual. Within this sphere are the beings of three ranks: Exusiai, Dynamis, and Kyriotetes.... The greater part of a human being's life between death and a new birth is spent among the Exusiai, Dynamis, and Kyriotetes.... In the Sun existence...there are only moral relationships. Everything moral in that sphere has the power of coming to realization in an appropriate way. Goodness produces phenomena that bring blessing to human beings, whereas evil brings the opposite.... In the Sun region moral relationships become reality. In this region every good intention, however feeble the thought, begins to be reality—a reality perceived by the Exusiai, Dynamis, and Kyriotetes. Human beings are regarded by the beings of the Sun region according to the goodness they had in thenselves, according to the way they were able to think, feel, and experience.... If, as a human being on Earth you have had good thoughts, in the Sun region between death and a new birth you will have converse with Exusiai, Dynamis, and Kyriotetes.[191]

The Outer Planets

There still remains the question: What takes place in the realm beyond the planetary spheres, in the realm of the zodiac, in the life after death? As mentioned earlier, the period in life after the age of 63 reflects our life in the zodiacal realm prior to incarnating upon the

Earth. The period in the zodiacal realm is of extraordinary significance, astrologically speaking. However, before discussing this, it is important to clarify something that often leads to confusion. This has to do with the newly discovered planets Uranus, Neptune, and Pluto—discovered in 1781, 1846, and 1930. One might think that Uranus, Neptune, and Pluto should be added into the sequence of seven-year periods as follows: Uranus (63–70), Neptune (70–77), Pluto (77–84), thus continuing the planetary periods up to the age of 84 (instead of 63). Why is this not the case?

First, as Rudolf Steiner pointed out, "Experiences...between the ages of fifty-six and sixty-three can be attributed to the influences of the Saturn sphere...When we have reached the Saturn sphere, we have passed through all that the beings of the planetary spheres can of their bounty accomplish for us. Then, in the esoteric sense, we embark upon a free and independent cosmic existence that looks back upon the planetary life."[192]

Here it is clear that after the age of sixty-three we proceed beyond the influences of the planetary spheres. We enter into the sphere of influence of the fixed stars (zodiac) beyond the planetary spheres that extend up from the Earth through seven planetary realms: Moon, Mercury, Venus, Sun, Mars, Jupiter, Saturn. Concerning the outer planets, this was Rudolf Steiner's indication for Uranus and Neptune, and the same obviously applies to Pluto, which at the time of Rudolf Steiner's comment had not been discovered: "We see that the two planets [Uranus and Neptune], which our modern astronomy sets on a par with the other planets, have in fact quite another origin and have fundamentally nothing to do with our universe." [193]

The planets Uranus, Neptune, and Pluto—as is clear from the astrological reincarnation research presented in this book—however do have an astrological significance. As described in *Hermetic Astrology*, vol. II, they work in a *transcendental* way, or in a subterranean way, but they do not work in the human being in the way the classical planets do. Thus, there are no internal organs corresponding to these three outermost planets. Also there are no Uranus, Neptune, and Pluto planetary spheres to be traversed after leaving the Saturn sphere in which we are helped to inscribe our Saturn karma with the help of the Seraphim. Therefore there are no seven-year periods in life corresponding to Uranus, Neptune, and Pluto.[194]

The Zodiacal Realm

After leaving the Saturn sphere and entering the zodiacal realm, a kind of "swoon" sets in, for which reason there is generally no recollection of the meeting with the Father, except in the case of the great initiates, such as the Master Jesus, known as the "Friend of God." On account of his ability to maintain consciousness through all the planetary spheres and also in the zodiacal sphere, the Master Jesus, in his pre-Christian incarnation as Zoroaster, was able to define the zodiac—that is, he was able to communicate the spheres of influence (sidereal signs) of the beings of the first hierarchy working through the zodiacal constellations.[195] It is only through the great initiates, or through higher spiritual beings, that it is possible to know what the human being undergoes in the sphere of the zodiac in the life between death and a new birth.

It is in this sphere of the zodiac that the decision to return to incarnation upon the Earth is taken. Then there begins a "looking toward the future," after having been more "looking back to the past," upon the path of ascent through the planetary spheres. The first thing that is decided upon, in looking to the future earthly life, is the horoscope, which is chosen in advance. The choice of horoscope is influenced by the "backward gaze" to the configuration at the preceding death, and through this to the configuration at the preceding conception and birth, and beyond this to the configurations at birth and death from still earlier incarnations. The configuration at the preceding death is particularly significant for the "backward gaze" since it appears like a "birth horoscope into the spiritual world."

The Midnight Hour of Existence

The human being, having made the decision in the zodiacal realm to reincarnate, chooses the future birth horoscope in harmony with the horoscope of death from the preceding incarnation. This is the background to the astrological "laws" of reincarnation, two of which are described in chapter 1 of this book.

A key element in the choice of the future birth horoscope is the *Ascendant*. It is the Ascendant that brings to expression the nature of the relationship that the human being has to the zodiac while indwelling the zodiacal sphere. It signifies the human being's basic orientation toward the encircling zodiac in which he finds himself to be spread out at the "midnight-hour of existence" between two incarnations. With the help of the Seraphim, Cherubim, and Thrones, who enjoy "the immediate sight of the Godhead," the human being builds up the spiritual archetype for the future physical body, drawing the forces for the head from the sidereal sign of Aries, for the larynx from the sign of Taurus, and so on through all twelve signs, finishing with the sign of Pisces as the constellation from which the forces for the feet are drawn. In the following description, the spiritual archetype of the physical body—generally known as the "spirit seed"—is called the "spiritual germ":

> A moment comes...*the great midnight hour of existence*...that brings to us this feeling: We must again become human beings. What we carried into the cosmos must be returned to us by the cosmos, so that we may come back to Earth.... Together with the spirits of the higher hierarchies and incorporeal human beings...we lay down the spiritual germ of the earthly human being. This is the great mystery: that the human being's heavenly occupation consists in weaving, in cooperation with the spirits of the higher hierarchies, the great spiritual germ of the future terrestrial human being. Inside the spiritual cosmos, all of us are weaving, in magnificent spiritual grandeur, the woof and weft of our own earthly existence, which will be attained by us after descending again into earthly life.... Not without justification did the ancient mysteries call the human physical body a temple.[196]

Macrocosmic Human Being

The spiritual archetype ("spiritual germ") of the physical body was depicted in ancient times as the human being in a circle around the zodiac, with the head in the constellation of Aries, the bright star Regulus (at 5° Leo, the Lion's heart) marking the heart, the feet in the constellation of Pisces, etc., thus indicating pictorially the traditional astrological correspondences of the parts of the physical body with the twelve signs of the zodiac.

> In the ancient mysteries the zodiac was represented pictorially. The signs of the zodiac are usually not depicted as animal forms. Each sign also represents a particular part of the human being, so that the microcosmic human being is depicted in this way: the head in the Ram, the region of the throat in the Bull, the arms expressing symmetry in the Twins, the cage of the ribs in Cancer, the heart in the Lion, and so on, until one reaches the lower part of the legs in Aquarius and the feet in the Fishes. Imagine the zodiac as the form of the human being inscribed in the cosmos. Then you will have a picture of the rudimentary parts of the human physical body as they are created in the midnight hour of existence by the power of the Thrones, Seraphim and Cherubim. This is the macrocosmic human being found in the myths of the world throughout the ages and out of which each individual human being has been composed in the most varied forms.... A profound truth lies at the basis of such representations; it comes to light more or less imperfectly depending on the clairvoyant capacities of the various peoples. It also shines through in the wisdom that found external expression in the Old Testament. We find it in the esoteric knowledge of the ancient Hebrews, which builds the foundation for the wisdom of the Old Testament: in the Adam Cadmon of the Kabbala. The macrocosmic human being is spread out in the cosmos.[197]

The pictorial representation of "Adam Cadmon" depicts the macrocosmic aspect of the physical body, i.e., the spirit seed or spiritual archetype of the physical body corresponding to the zodiac. This correspondence between the parts of the physical body and the twelve zodiacal signs is based on the spiritual reality of what we experience in the zodiacal realm during the midnight hour of existence, the midpoint between two incarnations. The Ascendant in the horoscope indicates the particular orientation that the human being takes up when in the zodiacal realm, and thus is significant as an indicator of the nature of the physical body.

Descent through the Planetary Spheres

After completing the work of building up the spirit seed ("spiritual germ"), the human being bears this spirit seed out of the zodiacal realm into the planetary spheres, beginning the path of descent back toward the Earth, commencing in the Saturn sphere. On the descent through the planetary spheres the *astral body* is built up from the seven planetary realms, incorporating the karmic inscriptions made in each realm.

It is usually the case, in the process of descent and building up the astral body, that the human being spends a longer period of time in one particular planetary sphere, thus coloring the structure of the astral body through the preponderance of the influence of that sphere. Here lies the background to the astrological teaching of the "ruling planet," which expresses the person's planetary type. As an example let us recall that between his incarnations as Raphael and Novalis, he encountered in the Jupiter sphere the one who later incarnated as Goethe:

> The Jupiter beings [Cherubim]...are, so to speak, wisdom that has become real and actual. Wisdom is innate in them; they cannot be other than wise. Just as we have circulating blood, so have the Jupiter beings wisdom. It is their very nature. Among them, too, karma can be shaped. The individuality of whom we are speaking, who lived through one of his most important lives in ancient Greece, passed through the Jupiter sphere, came into contact with the wisdom of the Jupiter sphere where his karma was shaped, and was born again in the eighteenth century as Goethe.[198]

Goethe was a Jupiter type. His "ruling planet" was Jupiter. Everything he wrote—his poems, novels, plays, especially his master work *Faust*—is inherently wise, permeated with the wisdom of the Jupiter sphere, which had absorbed into him through his lengthy sojourn in that sphere before descending into incarnation. One can have the question whether Goethe sensed his affinity to Jupiter when we read his words in the novel *Wilhelm Meisters Wanderjahre (Wilhelm Meister's Years of Travel)*:

With these words or thoughts he turned around, looking about, when he noticed...

> Jupiter, the star of fortune, more gloriously radiant than ever. He interpreted this as a good omen and remained joyfully in this beholding for a while.[199]

Here is not the place to enter into the complex question as to how and whether the ruling planet can be determined from the horoscope. It suffices to say that there is no simple astrological rule for determining the ruling planet. In Goethe's horoscope, for example, Jupiter is not particularly strongly placed. He was born with Saturn rising, and thus, Saturn is strongly placed. Also he was born with the Sun culminating, and so the Sun is also strong. One would expect either Saturn or the Sun to be the ruling planet in this horoscope. However, Jupiter is in the sign of Pisces, which is its "home constellation," and through the opposition of Venus with Jupiter it is activated "poetically" (recalling the connection of Venus to poets).

The main point here is that the human being, bearing the spirit seed, builds up the astral body on the path of descent through the planetary spheres, finally entering the Moon sphere. From the Moon sphere the incarnating human being influences (inspires) the future parents to come together to provide the physical body needed for the approaching incarnation. When the egg is fertilized in the mother's womb, the human being sends down the spirit seed from the Moon sphere to unite with the fertilized egg. "The spiritual germ woven by ourselves is lost at the moment when the physical germ, which we shall have to assume on Earth, is engendered through the act of conception."[200] In ancient times the sending down of the spirit seed to unite with the fertilized egg was seen clairvoyantly as the "descent of the stork," since the spirit seed appears to clairvoyant perception like a stork. When the stork was seen, it was known that a child was on its way.

The Epoch Chart

In terms of the epoch (conception) chart, the moment of the sending down of the spirit seed is computed using the hermetic rule retrogressively from the birth chart, utilizing the interchange of the position of the Moon and the Ascendant or its opposite.[201] Hermetic astrology, based on the application of the hermetic rule, consciously includes the moment of conception (epoch) as an important part of the process of incarnation. At this moment there begins the building up of the etheric body from the cosmic ether, and the weaving of the "karmic cross" into the etheric body, as described in detail in *Hermetic Astrology*, vol. II.

The epoch chart gives a picture of the structure of the etheric body, containing the basic life force. Research shows that the moment of death is often indicated by *transits* to the epoch chart. This discovery highlights the significance of the epoch chart in hermetic astrology. For example, as already referred to in the case of Novalis:

> *Death g-Saturn* (25½° Cancer)—*conjunct*—
> *Epoch Sun* (25½° Cancer)

And in the case of Raphael:

> *Death ♄-Saturn* (9° Capricorn)—*opposite*—
> *Epoch Sun* (8½° Cancer)
> *Death Sun* (8° Aries)—*opposite*—
> *Epoch ♄-Saturn* (7½° Libra)

And in the case of Rudolf Steiner;

> *Death g-Saturn* (19½° Libra)—*conjunct*—
> *Epoch Moon* (19½° Libra)[202]

These are just a few examples illustrating the research finding that the moment of death is often indicated by transits to the epoch/conception chart.

All this goes to show that the conception (epoch) chart, which is central to hermetic astrology, is highly significant. It has been largely neglected in the history of astrology, although working with the conception chart was a part of ancient Egyptian and Babylonian astrology. As Ptolemy states in the *Tetrabiblos*:

> Since the chronological starting-point of human nativities is naturally the very time of conception, but potentially and

accidentally the moment of birth, in cases in which the very time of conception is known either by chance or by observation, it is more fitting that we should follow it in determining the special nature of body and soul, examining the effective power of the configuration of the stars at that time.[203]

One of the reasons for the neglect of the conception horoscope in astrology is that the application of the hermetic rule to find the epoch from the birth chart is complex and open-ended. Now, through the Astrofire computer program developed by Peter Treadgold, the various possibilities for the epoch can be seen at a glance, and with a little practice and experience it is relatively easy to determine the actual conception horoscope.[204] This opens—or rather re-opens—a vast new realm of research intrinsic to the roots of astrology.

Against the cosmic background of the soul's journey through the planetary spheres in the life after death, it can be seen that the moment of the epoch is the time when the spirit seed is sent down from the Moon sphere to unite with the fertilized egg in the mother's womb. At this moment the soul feels a great sense of loss, amounting to bereavement.

The spiritual germ of the physical body has already descended to Earth, whereas we still dwell in the spiritual world. And now a vehement feeling of bereavement sets in. We have lost the spiritual germ of the physical body. This has already arrived below and united itself with the last of those successive generations, which we have watched. We ourselves, however, are still above. The feeling of bereavement becomes violent. And now this feeling of bereavement draws out of the universe the needful ingredients of the world ether. Having sent the spiritual germ of the physical body down to Earth and remained behind...we draw etheric substance out of the world ether and form our own etheric body.[205]

Out of this sense of loss begins the weaving of the etheric body from the cosmic ether, parallel to the formation of the physical body (embryo) in the mother's womb. This is the key significance of the epoch chart: it indicates the start of life, the beginning of the formation of both the physical body (through the spirit seed) and the etheric body (in relation to the Moon's orbit of the sidereal zodiac). Against this background it can be understood why *transits* to the epoch chart might be significant at the moment of death, which is the moment the etheric body withdraws from the physical body. In the course of further research into these hitherto hidden and mysterious connections, it became evident that the two epoch charts (hermetic and geocentric) relate to the physical and etheric bodies, respectively. At conception the hermetic chart—related to the level of the Sun—gives a picture of the *spirit seed* of the *physical body*, and the geocentric chart gives a picture of the *primary impulse* from the surrounding etheric cosmos on the building up of the *etheric body*.

Similarly the two birth charts (hermetic and geocentric) relate to the "I" and astral body of the human being. At birth the hermetic chart, which is connected with the level of the Sun, reveals the *goal* of the incarnating human being as an *"I"* coming into a new incarnation upon the Earth in order to fulfill a destiny elaborated together with the beings of the spiritual hierarchies upon the descent from the zodiac through the planetary spheres. And the geocentric chart at birth, which is the traditional astrological chart, has always been called the *"map of the soul"*—this clearly relating to the level of the *astral body*, which is the bearer of the soul (sometimes called the *soul body*).

Against this background it is evident that every human being has four horoscopes, relating to the four levels of the human being:

Hermetic conception horoscope:	physical body
Geocentric conception horoscope:	etheric body
Geocentric birth horoscope:	astral body
Hermetic birth horoscope:	"I"

The "Why" of Astrology

Contemplation of the soul's journey through the cosmic world in the life after death is an awe-inspiring story. It reveals to us the "why" of astrology. It points us to our celestial origin. It educates us to be better people, morally and spiritually. It re-connects us with our source. It opens up to us the possibility of becoming "citizens of the cosmos" in a conscious way. It counterbalances a one-sided perspective of reincarnation—the perspective that focuses solely upon the sequence of earthly lives.

In further service to this "homage to Elijah" contemplating the sequence of lives: Phinehas, Elijah, John the Baptist, St. Elisabeth of Hungary, Raphael, Novalis,[206] it was essential also to look at the period between incarnations, at least in between the incarnations as Raphael and that as Novalis. Since Raphael died in 1520 and Novalis was born in 1772, the midpoint ("midnight hour of existence") was in 1646, just prior to the conclusion of the Thirty Years' War in 1648. With the end of the Thirty Years' War, or around this time, this individuality made the decision to reincarnate upon the Earth to help bring a new culture after the destruction and havoc wreaked by this war that tore Europe apart.

Together with other great spirits (Goethe, Schiller, Beethoven, Lessing, Herder, Jean Paul, Friedrich Schlegel, August Wilhelm Schlegel, Fichte, Hölderlin, Schelling, Schleiermacher, Ludwig Tieck—just to name a few), Novalis incarnated to help in the creation of a new Europe. He became the pioneer of the German Romantic Movement. This was a movement from the purely intellectual thinking of his time toward "thinking with the heart." This is the activating principle for the faculty of "spiritual touch"—requiring the activity of the imagination to reach toward the underlying truths of super-sensible realities. Mysticism was, for Novalis, a natural response to the creative pulse of inspiration, and he used the word *romanticize* to characterize the activating principle of spiritual imagination:

> The world must be romanticized. In this way one discovers it's original meaning. Romanticizing is nothing but a qualitative raising to a higher power. In this operation the lower self becomes identified with the higher self. Just as we ourselves are a qualitative exponential series. This operation is still wholly unknown.[207]

Considering the soul's journey through the planetary spheres between death and rebirth, our perspective of Novalis is enriched, knowing of the meeting between the individuality of Novalis and that of Goethe in the Jupiter sphere prior to incarnation. Perhaps it was through this meeting that Novalis could understand Goethe's *Fairytale* as revealing the "Veil of the Eternal Virgin," and see in its central character, the Beautiful Lily, the "ideal human within humanity." Novalis referred to Goethe's *Fairytale* as the "canon of poetry," "a process of transmutation, purification, and reduction."[208]

What is the mystery underlying the relationship between Novalis and Sophie von Kühn, whose death birthed his mystical longing? Who was Sophie von Kühn? When Goethe heard of Sophie's illness, he visited her and was deeply moved at the sight of her suffering. Novalis wrote of this meeting:

> I shall not conceal from you that I could not regard him (Goethe) as the apostle of beauty, had he not been moved at the very sight of her. Surely it is not passion. It is too inexorable, too cold, too deeply rooted in my soul—this feeling that she is one of the noblest, idealest figures ever to walk the Earth.[209]

Novalis dedicated his *Hymns to the Night* to Sophie von Kühn, his immortal beloved. The work was intended as an expression of his experience of the *transfiguration* of his earthly beloved leading to his connection to the Eternal Feminine. And it was nothing less than an *experience* of the wisdom of the Eternal Feminine that inspired the birth of his own poetic genius.

A Seed-Bearing Work toward the Future

"Friends, the ground is poor; we must strew abundant seed that we might reap a modest harvest."[210] This work is intended as an introduction to the new astrology arising from astrological reincarnation research, and obviously not all questions can be answered in an introductory work. Be that as it may, hopefully you have been provided a glimpse into an exciting new world, one that concerns the higher calling, nature, purpose, and destiny of every human being. With this "homage to Elijah" we have followed the application of hermetic astrology, as a science of reincarnation and karma, to one of the great leaders of humanity—Elijah. From this exploration we can recognize the Elijah stream of incarnations as the spiritual impulse to bear witness to the living Word. It is a call for morality. It is a voice of conscience calling out in the most arid landscapes of the human soul as a herald for the "I" nature born from the spiritual Sun:

Conscience "is the innate mediator of every human being. It takes the place of God on Earth."[211]

15

THE TWO WITNESSES

MOSES AND ELIJAH—THE TWO WITNESSES

I will grant my two witnesses power to prophesy for one thousand two hundred and sixty days, clothed in sackcloth. These
are the two olive trees and the two lampstands that stand before
the Lord of the Earth. (Revelation 11: 3–4)

RAPHAEL DEPICTED THE "TWO olive trees and the two lampstands standing before the Lord" in his great unfinished masterpiece, the *Transfiguration*. At the transfiguration on Mt. Tabor,
when Christ Jesus was transfigured into radiant light, the three disciples—Peter, James, and John—beheld Elijah and Moses together
with the transfigured Christ. In Raphael's painting of the transfiguration, Moses and Elijah appear to the right and left of Jesus
Christ: Moses as the bearer of wisdom, and Elijah as the bringer
of strength. This presents us with a picture of the "two witnesses"
referred to in the Book of Revelation:

If anyone would harm them, fire pours from their mouth and
consumes their foes…They have power to shut the sky, that
no rain may fall during the days of their prophesying, and they
have power over the waters to turn them into blood, and to
smite the Earth with every plague. (Revelation 11:5–6)

From this description we recognize Moses and Elijah.[212] We
remember Elijah's words to the captain of fifty of the enemy: "Let
fire come down from heaven and consume you and your fifty"

(II Kings 1:10). Further, "Elijah said to Ahab: As the Lord God of Israel lives, before whom I stand, there shall be neither dew nor rain these years, except by my word" (I Kings 17:1). Thus, Elijah is identified as a prophet who calls forth the fire to pour down and consume his foes, also having the "power to shut the sky, that no rain may fall."

Similarly, Moses is identified as one of the two witnesses by the following words: "He lifted up the rod and struck the water that was in the Nile, and all the water that was in the Nile turned to blood" (Exodus 7:20). Further:

> The Lord said to Moses, "Go to Pharaoh and say to him, Thus says the Lord, the God of the Hebrews: Let my people go, that they may serve me. For if you refuse to let them go and still hold them, behold, the hand of the Lord will fall with a very severe plague upon your cattle." (Exodus 9:1–3)

This was one of numerous plagues with which the Earth—or rather the land of Egypt—was smitten through the command of Moses:

> And there has not arisen a prophet since in Israel like Moses, whom the Lord knew face to face, none like him for all the signs and wonders that the Lord sent him to do in the land of Egypt. (Deuteronomy 34:10–11)

As a witness to the glory of God, the Zohar separates Moses by many degrees from all the other prophets:

> Moses raised his head without fear and gazed at the brightness of the supernal glory without losing his senses...even when he was in that exalted grade he did not lose his faculties, and straightway after gazing on the brightness of the heavenly glory he "returned to camp" to speak to them concerning all require-ments, and his mind was as clear as before, and more so.[213]

This is the meaning of "face to face" in Old Testament times—remain-ing clear and conscious in expanded states of consciousness.

We can thus recognize the individualities of Moses and Elijah as the two witnesses, for their stories bear within themselves a responsibility toward the moral development of humanity—toward the future unfolding of the evolution of consciousness.

Toward the Unfolding of the Twenty-first Century

The unfolding of spiritual impulses in the twentieth century and on into the twenty-first century can be understood against the background of the working of the two witnesses, Moses and Elijah. The guiding inspiration for my book *The Most Holy Trinosophia and the New Revelation of the Divine Feminine* was to work toward an understanding of Rudolf Steiner's prophecy that the reincarnated Elijah (as the reincarnated poet Novalis) would reincarnate toward the end of the twentieth century,

> to prepare the work that shall be accomplished at the end of the century and that shall lead humankind past the crisis in which it is involved.[214]

This book is a further *homage* to Elijah and my hope is that it may serve as a stimulus to enable an inner connection to be forged with this great leader of humanity, whose task is to lead humankind through the present crisis of materialism in which human beings are currently embroiled. An awakening (or re-awakening) to a conscious relationship with the starry heavens is an important aspect in the overcoming of materialism, as this facilitates an awakening to one's true self.

> The star world now is flowing,
> As living, golden wine,
> Its joys on us bestowing,
> Ourselves as stars will shine.[215]

The Second Witness:
Prophet for the Coming of the Rose of the World

As discussed in my book *The Most Holy Trinosophia*, the promise of the return of the witnessing stream of the Elijah–John individuality in our time is taking place in a female incarnation. This female incarnation can be regarded as that of the third teacher (in a series of incarnations of three spiritual teachers)—the first representing the Archangel Michael, the second representing the Etheric Christ, and the third representing Divine Sophia. Consideration of the reincarnated Elijah individuality as the third teacher is the deeper significance of the title *Elijah Come Again.*

The present incarnation of the Elijah individuality was prophesied in 1924 by Rudolf Steiner in the last lecture he held.[216] Although he did not speak in this lecture of this as a female embodiment, his prophecy of the return of Elijah does help to illuminate the prophecy in chapter 2 of the Book of Revelation concerning the coming of the two witnesses, one of whom is Elijah. If we consider the individuality of Elijah as the second witness for the second coming of Christ, then the individuality of Moses would have preceded Elijah, serving in the early part of the twentieth century as the first witness, as elaborated upon in my earlier book, *The Most Holy Trinosophia.*

The message of this book *Elijah Come Again* is that the return of the Elijah individuality as the second witness for the second coming of Christ, is an event occurring in *our* time, with this incarnation coming in the latter part of the twentieth century and the early part of the twenty-first century. This event, the coming of the Elijah individuality as a *woman* bearing witness to the second coming of Christ, is *simultaneously* a heralding for the future coming of Divine Sophia, as the Rose of the World in the future Age of Aquarius.[217] It has to be borne in mind that the female incarnation of the Elijah individuality in our time is one that is wholly behind the scenes, not in any way in the public arena. This book has been written to help the reader find an inner connection with this reincarnated Elijah individuality, whose work is now unfolding, and to connect with her activity as a real presence and force in today's world—the same presence and force as the disciples experienced after the beheading of John the Baptist, when John worked inspirationally as a kind of "group soul" overlighting the circle of Christ's disciples.

As referred to earlier, the two witnesses—Moses and Elijah—are represented in Raphael's great work *The Transfiguration.* Moses appears at the right-hand side of the transfigured Christ; and Elijah appears at His left side (seen to the right from the viewer's perspective). As an unfinished masterpiece, Raphael's painting, serves also as a foretelling of what is yet to come. *The Transfiguration* can be regarded as a prophetic heralding of the Second Coming of Christ and as a portrayal—at the time of the Second Coming—of Christ working together with the two witnesses: Moses as the "right-hand man" of Christ, and Elijah on the feminine (left-hand) side of Christ.

The second witness, the reincarnated Elijah individuality, serving now in a female form, as the *bearer of strength*, is bearing the strength of the feminine side of Christ. This can be understood in connection with Divine Sophia, as Sophia is the feminine counterpart of Christ.

An archetypal example of strength as a bearer of the Divine Feminine can be found in the eleventh Tarot card—entitled *Strength*—which appears on a powerful female figure whose strength is attained through purity.[218] The deeper message of this image from the major arcana of the Tarot is that the strength of the human will lies in its alignment with Divine Will, and this alignment is possible only through achieving a certain level of purity.

There is yet another aspect to the coming of the two witnesses as heralds of the second coming of Christ and of the coming of Divine Sophia, and that is as *speakers* on behalf of the spiritual world in the face of the coming of the Antichrist. The advent of the Antichrist in our time is a theme that is too vast to be entered into here.[219] It suffices to say that, as witnesses of the Second Coming of Christ, the two witnesses suffer the wrath of the Antichrist, also known as *The Beast*.

> And when the time of their witnessing is fulfilled, the beast that rose from the abyss will fight them and defeat them and kill them. (Revelation 11:7–8)

The *Revelation to John* goes on to describe how the two witnesses rise from the dead and ascend to heaven. In other words, they defy the death-bringing forces of the Antichrist and are the subjects of miracles that could be described as *divine magic*. These are images that need not necessarily be taken literally. The important point is that at this time of the coming of the Antichrist, Elijah, as one of the two witnesses, is incarnated and offers a powerful stream of inspiration to all who turn to her at this time.

Divine Sophia—The Rose of the World

Holding in consciousness the sequence of incarnations of this great individuality—Phinehas, Elijah, John the Baptist, Elisabeth of Thüringia, Raphael, Novalis—together with the contemplation of Raphael's painting *The Transfiguration*, depicting Christ centered between his two witnesses Moses and Elijah, the latter as a bearer of strength appearing to the left of Christ (the receptive, feminine side), facilitates an understanding of the future unfoldment of the feminine aspect of Christ. Contemplation of this powerful image against the background of all that has been set forth in the preceding pages can serve as a doorway or opening to the flow of inspiration now proceeding from the Elijah individuality as the second witness for the Second Coming of Christ—"The radiant and splendid forerunner...who shall lead humanity past the great crisis in which it is involved."[220]

These mysteries can also be accessed through a contemplation of the Russian icons depicting Divine Sophia. These icons generally depict Sophia seated on a throne with the Virgin Mary and John the Baptist on either side. This depiction is a telling image of Mary's and John's connection to Sophia. Further, we find in the archetypal icons of the Divine Sophia a key to understanding the spiritual task of the Elijah–John individuality in relation to Sophia in the new millennium—which has been called the "Sophianic millennium."[221] In the archetypal icon of Divine Sophia, the radiant figure of Sophia appears enthroned in the center with the Virgin Mary to the left of the throne (seen from the viewer's perspective) and John the Baptist on the right.[222] As Mary and John both walked the Earth in human incarnations as messengers on behalf of Christ two thousand years ago, so now they serve as messengers or mediators for Sophia proclaiming a new era to humanity.[223]

Appearing as a herald for the advent of the Sophianic millennium around the world, there have been—and continue to be—numerous apparitions of Mary. As herald, Mary serves as a messenger of Sophia, mediating between Sophia and humanity. Mary, as mediator, comes to expression in Teilard de Chardin's *Hymn to the Eternal Feminine*:

I am the unfading beauty of the times to come—the Ideal
    Feminine....
It is God who awaits you and me!
He caused me to rise up; a luminous mist over the abyss—
Between the Earth and Himself—that, in me, he might dwell
    among you...
I am Mary the Virgin, Mother of all humankind...
I am the Eternal Feminine.

If we consider that the icon of Divine Sophia is a representation of spiritual reality, it is clear that with the increasing number of Marian apparitions bearing witness to Sophia's approach to humanity the Elijah–John individuality is also becoming active in an inspirational way, as a source of inspiration towards an increasing consciousness of Divine Sophia. We see this interweaving especially in the arts coming to expression in the great Madonna paintings of Raphael and in the poetry of Novalis. Sophia's inspiring influence permeates the cosmology of Novalis. As one student of the works of Novalis concludes, "His cosmology is based on the insight that the human soul is an image of the whole cosmos."[224] Such is the keynote of a cosmological view inspired by Sophia, who is called "the Soul of the World."[225]

Through the words of Novalis we can experience that the Elijah individuality is also at work in an inspirational way toward the development of a new wisdom of the stars—Astro-Sophia (Astrosophy)—dedicated to unveiling the cosmic mysteries of Divine Sophia, which is the foundation and the *raison d'être* of this book.

"The Transfiguration"—an Unfinished Masterpiece

"Every vision contains some mystery until its fulfillment."[226] These words, received interiorly by Anne Catherine Emmerich can help in our contemplation of Raphael's painting *The Transfiguration*, for the painting appears as a great vision of the transfigured Christ appearing as if floating in an aura of light and clouds. Later Novalis was to write of Christ's future coming. In the twelfth of his *Sacred Songs* the poet describes the "World-wide Comforter" *as if appearing out of the clouds*:

O drive him from Thee to our arm
While round Him Thy breath is warm;
Then gather heavy clouds to sheath Him
And let Him come to Earth beneath them.

In cooling streams now send Him here,
In flames of fire let Him appear,
In air and oil, in sound and dew
The Earth's foundations piercing through.

The holy battle thus is fought,
The wrath of Hell is turned to naught,
And flowering everywhere around
Our former Paradise is found.

The Earth grows green and stirs to life,
While filled with Spirit all men strive
Their Savior lovingly to meet
And with o'er-flowing heart to greet.

Raphael's poor health prevented him from finishing painting *The Transfiguration*. Therefore his students (in a role such that they might have earlier been called "disciples") were called upon to complete the lower half of the painting, using Raphael's preliminary sketch depicting the disciples of Christ gesturing toward the light of the Christ's glory on the one hand and, on the other hand, pointing down toward the healing of the possessed boy, whose gesture of freedom points toward the exalted Christ.

Thus, the painting seems to hint toward humanity's responsibility to complete that which has been indicated in the depths of the painting, as an inspiration for the future. Modern humanity must now address the reality of possession by the extreme materialism of our time—with its myriad of illusory temptations that blind us to the reality of our spiritual connection to the cosmos.

How might our modern dilemma be seen in relation to the Elijah stream of incarnations? Specifically, how might the relationship of John–Raphael's life and works, which lead us to light-filled spiritual heights, be seen *together* with the Lazarus–John mystery, having to do with the transformation of the depths? The following commentary would indicate that this is an issue that is of great importance in approaching the future. Following a conversation that he had with Rudolf Steiner on March 3, 1925, Count Ludwig Polzer-Hoditz wrote:

> Rudolf Steiner's research into...the mystery of the merging of the two Johns, indicated in his last address, becomes a spiritual legacy that calls upon us to strive continually for an explanation. It is moreover a question, the solving of which is of particular importance for the future. This has been preserved as one of the last utterances of Rudolf Steiner's life.[227]

Perhaps this reveals yet another aspect connected to the twofold content of *The Transfiguration*, recalling that Raphael completed only the upper "light filled" half of the painting, whereas the lower half, depicting the healing of the possessed boy, remains as an inspiration for us to contemplate the human soul condition of our time, together with our individual responsibility toward the transformation of the depths. This is represented archetypally in the *uniting* of John the Baptist with Lazarus, who represent two different streams: the stream of Elijah–John leading to the light-filled spiritual heights, and the stream of Lazarus–John leading to the depths.

In her essay relating to the "merging of the two Johns," Hella Wiesberger first quotes from Rudolf Steiner's lecture about the profound significance of the red rose, as the symbol for the human soul that is developing personality, and the white lily that symbolizes the soul that remains spiritual insofar as the "I" largely remains in celestial heights, spiritually permeating the soul from above. In contemplating the working together of the soul-spiritual aspect of the "I" symbolized by the white lily, and the developing personality of the human soul symbolized by the red rose, how might we understand the following words of Rudolf Steiner? "In the uniting of the Lily soul and the Rose soul was envisaged what can unite with the Mystery of Golgotha" (lecture of May 6, 1909).

Hella Wiesberger develops this further:

> In how far the contrast of the Rose soul and the Lily soul can be applied to the two John individualities is shown by the fact

that...Lazarus is always characterized as the representative of the forces of personality, whereas the Elijah soul is often described as such a highly spiritual being that he can only be loosely connected with his earthly vessels, as was also the case with John the Baptist. If the uniting of the Rose soul and the Lily soul can lead to union with the Mystery of Golgotha, we may conclude—in view of the merging of the two John souls at the raising of Lazarus by Jesus Christ—that the disciple whom the Lord loved has *become* the being to whom the Christ secret of the overcoming of death has become attached.[228]

Further light is shed upon the mystery of the two complementary "John streams" by Johanna von Keyserlingk. Following a last lecture on September 28, 1924—the last lecture given before his death in March, 1925—it is reported by Johanna von Keyserlingk that Rudolf Steiner said in a conversation with his medical doctor, Ita Wegman, that he had "created a new stream." According to Johanna von Keyserlingk, what Rudolf Steiner meant by this is that he had created a stream that "leads into the depths."[229] She indicates, on the basis of her esoteric discussions with Rudolf Steiner, and also derived from knowledge that he had communicated on other occasions, that "the spiritual paths of John–Raphael lead into the Light regions of the Buddha-sphere." On the other hand, she says, "the spiritual paths of Lazarus–John lead to the...mysteries of the Earth depths." She points out that the latter path has to do with the mystery of Holy Saturday, when Christ descended into the depths of the Earth, in order to redeem the interior of the Earth, including the Cain region near the center of the Earth. She refers to Rudolf Steiner's indication that Lazarus–John was the reincarnated Cain,[230] "who has his dominion in the inner Earth," and she goes on to say: "The Holy Saturday mysteries that are still hidden in the Earth are at the same time the Saturn mysteries of Lazarus–Cain."[231]

Was this the mystery that led Novalis into the depths of the Earth as a mining inspector in the region of Thüringia, which he equated to mining the depths of the soul, a noble occupation to be measured with morality and wisdom seeking? Might Rudolf Steiner's uniting of the light-bearing stream and the stream leading into the depths be an indication of the task at hand for humankind—the healing of the possessed boy in the depths, the story of possession in the depths of our own soul—seeking regeneration through the inspiring force of Elijah come again!

At the close of this last lecture after discussing the sequence of incarnations: Elijah-John the Baptist–Raphael–Novalis, Rudolf Steiner spoke of this individuality as, "A radiant and splendid forerunner...with whom you are to prepare the work that shall be accomplished at the end of the [twentieth] century, and that shall lead humankind past the great crisis in which it is involved."[232] These words indicate that, from the end of the twentieth century and into the twenty-first century—that is, *now*—the Elijah–John individuality is to be the "bearer of strength" for us in the next step *toward our spiritual evolution*. Perhaps the mystery of *The Transfiguration* having been left unfinished is that this is a masterpiece that includes the Earth and all of humanity.

> Where art thou; world-wide comforter?
> Thy place is long prepared here.
> Each opens his soul to Thee,
> And awaits Thy blessing, yearningly. [233]

AFTERWORD:
FULFILLMENT OF A PROPHECY

L ET US RECALL RUDOLF Steiner's words concerning the individuality of Novalis from his lecture in September, 1924:

> And so we see in Novalis a radiant and splendid forerunner
> of that Michael stream which is now to lead you all, my dear
> friends, while you live; and then, *after you have gone through
> the gate of death, you will find in the spiritual suprasensory
> worlds all those others—among them also the being of whom I
> have been speaking to you today*—all those with whom you are
> to prepare the work...that shall lead humankind past the great
> crisis in which it is involved.[234]

With these words, addressed in 1924 to those around him, Rudolf
Steiner made it clear that at that time the individuality of Elijah–
John–Elisabeth–Raphael–Novalis ("the being of whom I have been
speaking to you today") was in spiritual realms, i.e., not in earthly
incarnation. Careful study of the whole context of this quote from
Rudolf Steiner's *Last Address* reveals a prediction concerning the
reincarnation of Novalis toward the end of the twentieth century,
since Steiner made it clear that all of those to whom he was speaking
would—after death—join the Novalis individuality in the spiritual
world and would then, under the leadership of the Novalis individuality, which implies "together with this individuality," reincarnate
in the latter part of the twentieth century to "lead humankind past
the great crisis in which it is involved." Keeping this in mind, here it
is appropriate to draw attention to the indication referred to in *The
Most Holy Trinosophia* of the prophecy of the reincarnation of the
Novalis individuality, the third teacher, in feminine form during the
latter part of the twentieth century:

It has been predicted (see chapter 3: The Holy Soul) that the third spiritual teacher of the twentieth century, who represents Sophia, would incarnate in female form.... This individuality is working especially strongly now...inspiring love and community among spiritually striving human beings.... It is the impulse of the Holy Soul that lives most strongly through the third teacher of the twentieth century.[235]

The prediction referred to here expands upon Rudolf Steiner's 1924 prediction in that it identifies the reincarnated Elijah–John–Elisabeth–Raphael–Novalis individuality as the third teacher of the twentieth (extending into the twenty-first) century, and also indicates the incarnation of this third teacher *in female form*. The incarnation of the third teacher as a female signifies the fulfillment of a prophecy made in 1933.[236] The identity of the Novalis individuality as the third teacher was published in *The Most Holy Trinosophia* in the year 2000, where also the timing of this incarnation is indicated.[237] As I wrote at that time:

The third teacher will almost certainly remain hidden, working behind the scenes as a source of inspiration, a vessel for Sophia, to strengthen wills in the trials of faith now coming upon the whole world.[238]

In *The Most Holy Trinosophia*, I indicated "around the year 1975" for the rebirth of Novalis in female form.[239] It is a matter of the fulfillment of the prophecy implicit in the title of this book *Elijah Come Again*, which has been written to help the reader find an inner connection with this reincarnated Elijah individuality, whose work is now unfolding as a blessing for humanity and the Earth.

As described in *Christ and the Maya Calendar: 2012 and the Coming of the Antichrist*,[240] the reincarnated Elijah–John–Elisabeth–Raphael–Novalis individuality is the third of three spiritual teachers, whose mission is to serve as teachers and guides of the community of those seeking Christ in his second coming, appearing one after the other in the course of time. "The work of the three teachers may be seen as bearing a direct relation to the stages of incarnation of the Etheric Christ, with the teachers acting as 'ambassadors of Christ' in the New Age."[241] After the first two Christ teachers of the twentieth century,[242] whose spiritual work centered around proclaiming (or even facilitating) Christ's return in the etheric realm, the third spiritual teacher is now unfolding her activity, while remaining hidden behind the scenes. In this respect, the work of the third teacher is a fulfillment of that of the two teachers before her, bringing the work of the Etheric Christ a stage further at this crucial time in the twenty-first century. To those who align themselves with her, she represents to the world the triumph of the Etheric Christ. Her presence on Earth points toward the fulfillment of the prophecy implicit in the title of this book, *Elijah Come Again.*

APPENDIX I:
TRUTH AND ERROR IN KARMA RESEARCH

IN PRESENTING THE ASTROLOGICAL reincarnation research for the Elijah individuality, I have drawn upon the use of the Astrofire computer program for the computing of the comparison of conception, birth, and death horoscopes of the same individual from incarnation to incarnation. Like the use of the telescope in astronomy, this computer program is a very powerful tool. However, whereas an astronomical telescope could not be thought of as dangerous, this new tool that has recently become available is unfortunately open to misuse, if the research *potential* that the computer program makes possible is not guided by the right principles. Therefore, in this appendix I would like to summarize some of the principles that need to be observed in applying this new research tool.

Right Use of Will

First and foremost it needs to be borne in mind that previous incarnations cannot be calculated simply by applying the astrological "laws" of reincarnation. The method of chart comparison described in this book has been applied solely to instances where the previous incarnations are already known. The aim has been to discover the laws and principles fundamental to astrology, not to discover previous incarnations. Any attempt to utilize a computer program for the purpose of "discovering" previous incarnations would be an inappropriate use of this tool. At most there can be a confirming application, where on some level, the previous incarnation is already known and the method of chart comparison is applied for confirmation and to discover the thread of continuity between the incarnations—the soul's highest intention, so to speak. In this case the effect of applying this method is that something is raised more into consciousness and is known in a more concrete and definite way.

Here, now, with an outline of some of the pitfalls belonging to research into reincarnation and karma. The major pitfall is that it can happen in doing *any* chart comparison that there are zodiacal alignments purely by chance. If such alignments are particularly striking, the person making the chart comparison might be tempted to believe that it is a matter of reincarnation. It is only through a long and subtle training—a whole process demanding an extraordinary degree of conscientiousness, combined with spiritual and moral development—that it is possible to arrive at any level of certainty as to whether two charts really "belong together" from one incarnation to the next. The possibilities for self-delusion here are unlimited. It is solely in the spirit of a quest for truth, and by testing one's findings over and over again, that one is able to learn in the course of time to distinguish a true reincarnation from an imagined one by the method of chart comparison.

Subconsciously there is in most people a longing for fame and acknowledgment.

If one cannot be famous in this life, perhaps one was somebody famous in a previous incarnation? This desire for fame, however, has to be overcome completely in order to arrive at truth in the realm of reincarnation and karma. Thus, humility is an essential practice on the spiritual path.

Illusions and False Inspirations

After death, the truth that one has been living in an illusion will inevitably be revealed. How could it happen that an enlightened soul could come to identify with some historical personality, and yet not be that personality? There are a number of striking examples of this. Such a false identification on the level of the self (ego) can be called an "illness of the self (ego)," for it creates a false sense of self, which is not grounded in reality. Usually this illness is discovered only after crossing the threshold of death, when one's spiritual eyes are opened. One of the purposes of this book is to awaken consciousness of the danger of this spiritual illness, which is now becoming more and more widespread. How is it possible that one could fall victim to a "false inspiration"? First, let us consider Rudolf Steiner's words:

> After you have gone through the gate of death, you will find in the spiritual suprasensory worlds...the being of whom I have been speaking to you today.[243]

Here Rudolf Steiner was speaking to people living in the year 1924, saying that after their death and transition to spiritual realms they would find the being of the individuality of Elijah–John the Baptist–Raphael–Novalis, "of whom I have been speaking to you today." Thus, Rudolf Steiner clearly states that the individuality whose last earthly life was that of Novalis was *not* incarnated in the year 1924, but could be found "in the spiritual, suprasensory worlds."

It is interesting to know that a significant individual of the twentieth century, who was a young person in 1924, believed himself to have been the reincarnated Novalis. Could it be that although he himself had not been the historical personality he thought he had been, in this case Novalis, that he was however spiritually connected with this being—and thus could feel the being of Novalis as a living source of inspiration in the twentieth century? In this way he came to identify himself with the source of his inspiration. But how could this happen? It is through the faculty that Rudolf Steiner calls *Intuition*, which can be likened to the sense of "spiritual touch," that it is possible to develop the faculty to be in spiritual realms, to tap into the Akasha Chronicle, and thereby to experience scenes and images from prior lives of others as if they were one's own inner experiences. This is one explanation of the phenomenon of "karmic memories" arising from another life that was not one's own previous incarnation.

Another example from a completely different perspective is that it is also possible that one could have observed the life of another from spiritual realms prior to one's own birth—from these realms beholding the earthly incarnation of another with whom one feels a spiritual connection. In a spiritual sense one could have been a "co-participator" in the life events of another, thus becoming the bearer of these "memories" during one's own incarnation. The memories of this other life arising would not be actual memories from earthly life—rather, they were reminiscences from one's own pre-incarnatory life, from an earlier spiritual vantage point looking down and beholding the earthly life of another. However, these spiritual reminiscences might appear to be actual recollections of one's own earthly life from a previous embodiment on the Earth.

Conclusion

These examples are characteristic for many people, as it is quite natural to feel a deep and powerful spiritual connection with the great leaders of humanity. In one's pre-incarnatory existence one beholds the lives of such individuals from the vantage point of the spiritual world. Their lives then later ray forth within one's own spiritual reminiscences. As Rudolf Steiner once remarked, in his lifetime he had met twenty-four reincarnated Mary Magdalenes![244] I must say that in my own experience I could add substantially to that number. In the case of some of these women it could well have been merely their wish and desire to have been "someone special," who had been so close to Christ. In other cases, however, it could have been genuine spiritual reminiscences of the life of Mary Magdalene beheld from spiritual realms that welled up from within as if they were actual memories of their own earlier incarnation on Earth.

In writing this, the intention is to reveal the subtle pitfalls connected with the "experience" of reincarnation. These experiences should not be taken simply at face value, but as something to be explored until their true meaning is revealed—just as we do not take dream experiences at face value but as something to be penetrated consciously. The same applies to "experiences" derived through hypnotic regression. These experiences are subject to question, since they are not from the self (ego) level of experience. It is possible, therefore, that they are not drawn from one's own subconscious but from the "collective unconscious." Hypnotic regression to discover one's "previous incarnations" is definitely not to be recommended, any more than the use of a computer program.

This discussion is by no means meant to discredit the "freedom" that can be gained under the guidance of a skilled and well qualified therapist toward breaking through outmoded and limiting belief patterns that could indeed be *echoes* from the past. Rather, the intention is to discourage an equally limiting possibility. In truth, our karmic task is to be in our *present* incarnation and to meet it with full responsibility and consciousness.

Against the background of these forewarnings, it can be understood why the traditional Christian churches reject reincarnation. Perhaps there is a deeper, esoteric reason why—in particular within the Catholic Church—the teaching of reincarnation was rejected.[245] One perspective concerning the reason for this rejection is that the danger exists that the human being could become so enamored with earthly life, and the sequence of earthly lives, that the greater life in spiritual realms between incarnations might be ignored.

Appendix 2: Astrological Reincarnation Chart Comparisons

Table of Astrological Symbols

Planets		Zodiacal Signs	
⊕	Earth	♈	Aries (Ram) AR
☉	Sun	♉	Taurus (Bull) TA
☽	Moon	♊	Gemini (Twins) GE
☿	Mercury	♋	Cancer (Crab) CN
♀	Venus	♌	Leo (Lion) LE
♂	Mars	♍	Virgo (Virgin) VI
♃	Jupiter	♎	Libra (Scales) LI
♄	Saturn	♏	Scorpio (Scorpion) SC
♅	Uranus	♐	Sagittarius (Archer) SG
♆	Neptune	♑	Capricorn (Goat) CP
♇	Pluto	♒	Aquarius (Waterman) AQ
		♓	Pisces (Fishes) PI

Categories:

The following horoscopes fall into two categories:

Section A: individual horoscopes—for events or for the conception, birth, and death of individuals.

Section B: comparison horoscopes—various comparison horoscopes, most of them relating to comparisons between the various incarnations of Elijah.

The Order of the Horoscopes

The same principle of ordering the horoscopes is used in Section A and in Section B. It is the principle of ordering the horoscopes chronologically. In Section A the chronological ordering is straightforward.

In the case of the comparison horoscopes in section B: they are ordered in the sequence of incarnations: first the comparison horoscopes relating to John the Baptist; then those relating to St. Elisabeth of Hungary; followed by those relating to Raphael; and finally those relating to Novalis. However, in the event of a comparison of a later incarnation with an earlier one, one has to look under the later one to find the comparison chart. For example, in the comparison of Novalis with John the Baptist, it is necessary to look under *Novalis*, and to observe the following guideline:

Each comparison horoscope comprises an inner circle and an outer circle, whereby the inner circle is the later horoscope, chronologically, and the outer circle is the earlier one. The chronological ordering of the comparison charts is organized in relation to the inner circle (indicating the later horoscope). Therefore, the very last comparison horoscope, chronologically, in this book is that between Rudolf Steiner (inner) and Thomas Aquinas (outer), since Rudolf Steiner, chronologically, follows Novalis (see pages 3 & 6).

Concerning the Zodiac

It should be noted that—with one exception—all zodiacal longitudes indicated in the text and presented in the following horoscopes are in terms of the sidereal zodiac, which has to be distinguished from the tropical zodiac in widespread use in contemporary astrology in the West. The one exception is the tropical chart comparison of the birth of Raphael with the birth of John the Baptist (see page 24). The Greek astronomer Claudius Ptolemy introduced the tropical zodiac into astrology in the middle of the second century C.E. Prior to this the sidereal zodiac was in use. Such was the influence of Ptolemy upon the western astrological tradition that the tropical zodiac became substituted for the sidereal zodiac used by the Babylonians, Egyptians, and early Greek astrologers. Yet Ptolemy did not influence the astrological tradition in India, and so the sidereal zodiac is still used to this day by Hindu astrologers.

The sidereal zodiac originated with the Babylonians in the sixth to fifth centuries B.C.E. and was defined by them in relation to certain bright stars. For example, Aldebaran ("the Bull's eye") is located in the middle of the sidereal sign/constellation of the Bull at 15° Taurus, and Antares ("the Scorpion's heart) is in the middle of the sidereal sign/ constellation of the Scorpion at 15° Scorpio. The sidereal signs, each thirty degrees long, coincide closely with the twelve astronomical zodi-acal constellations of the same name. Whereas the signs of the tropical zodiac, since they are defined in relation to the vernal point, now have little or no relationship to the corresponding zodiacal constellations. This is because the vernal point, the zodiacal location of the Sun on March 21, shifts slowly backward through the sidereal zodiac at a rate of one degree in seventy-two years ("the precession of the equinoxes").

When Ptolemy introduced the tropical zodiac into astrology, there was an almost exact coincidence between the tropical and the side-real zodiac, as the vernal point, which is defined to be 0° Aries in the tropical zodiac, was at 1° Aries in the sidereal zodiac in the middle of the second century C.E. Thus, there was only one degree difference between the two zodiacs. So, it made hardly any difference to Ptolemy or his contemporaries to use the tropical zodiac instead of the sidereal zodiac. But now—the vernal point, on account of precession, having shifted back from 1° Aries to 5° Pisces—there is a 25 degree difference and so there is virtually no correspondence between the two. Without going into further detail concerning the complex issue of the zodiac, as I have shown in my *Hermetic Astrology* trilogy, I remind the reader that it was the sidereal (according to the stars) zodiac that was used by the three Magi, who were the last representatives of the true star wisdom in antiquity.

A full and comprehensive study of the history of the zodiac, how it came to be defined in the first place, is the subject of my doctoral research, published in the book *History of the Zodiac*.

Concerning the Astrological Charts

Each chart indicates the positions in the sidereal zodiac of the Sun, Moon, Moon's nodes, and planets and also the Ascendant, the Midheaven, and the other astrological house cusps. The information relating to the Ascendant, the Midheaven, and the other astrological house cusps is relevant *only in cases where the time is known*, and can therefore be ignored if the substitute (average) time of midday is used.

(Note that Christ's ascension really did take place at midday, and so the time of noon used in this horoscope is the actual time of this event and not a substitute, average time.) The data is based on accurate computer calculations of the planetary positions, geocentric and heliocentric, at the time of the event or at the epoch/conception (∅), birth (✳) and death (†) of the person under consideration. These planetary positions are accurate if the time is known[246] and are only approximate if the substitute (average) time of noon is used. The approximate planetary positions are acceptable in all cases, with the exception of the Moon and heliocentric/hermetic Mercury, since the Moon's position can be up to 7½° different if the event took place around midnight, and it can be up to 3°10' different in the case of h-Mercury—both values exceeding the usual three degree range of tolerance for planetary alignments.

All planetary positions are listed in the sidereal zodiac to degree and minute; all dates are given in terms of the Julian or the Gregorian calendar according to which was in use at the time, and all times are given in the twenty-four-hour system either as local standard time or as zone standard time. (See the bibliography of works by Robert Powell for information concerning the astronomical basis of these charts.)

With grateful acknowledgment to Peter Treadgold for his Astrofire program (see details at the end of the Bibliography below), with which the charts and comparison charts for this body of research were produced. For anyone interested in working with or carrying out research in the field of astrosophy or hermetic astrology, this program is ideal, enabling one to produce charts such as the horoscopes in this book or such as the charts of Willi Sucher and others.

The original intention of the author and the publisher was to include in appendix 2 all the horoscopes and comparison horoscopes referred to in this book. However, as the number of horoscopes and comparison horoscopes amounts to about one hundred, this original intention has had to be abandoned. Instead, all the horoscopes and comparison horoscopes are to be found under "Elijah Come Again" on the website: www.astrogeographia.org, and a selection of these horoscopes and comparison horoscopes are published here in this book.

Horoscopes for Appendix 2

Section A:

Baptism of Jesus Christ - Geocentric
At River Jordon, near Ono, Latitude 31N50', Longitude 35E33'
Date: Friday, 23/SEP/29, Julian
Time: 10: 0, Local Time
Sidereal Time 10: 0: 1, Vernal Point 2 ♈ 38'42", House System: Placidus
Zodiac: Sidereal SVP, Aspect set: Conjunction/Square/Opposition

Chart 1: geocentric horoscope of the Baptism

Chart 2: heliocentric horoscope of the Baptism

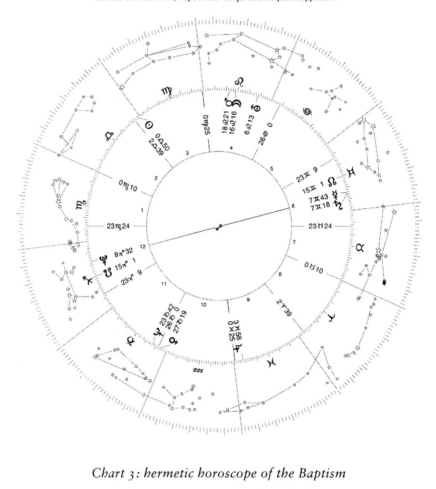

Chart 3: hermetic horoscope of the Baptism

Comparing charts 1, 2, and 3, it is only in the hermetic horoscope that the conjunction of the Moon and Mars at the Baptism appears—referred to as a "hermetic conjunction." Note that *hermetic* also means "hidden."

Ascension of Jesus Christ - Geocentric
At Mount of Olives, Latitude 31N46', Longitude 35E14'
Date: Thursday, 14/MAY/33, Julian
Time: 12: 0, Local Time
Sidereal Time 3:20: 3, Vernal Point 2♈35'41", House System: Placidus
Zodiac: Sidereal SVP, Aspect set: Conjunction/Square/Opposition

Chart 4: geocentric horoscope of the Ascension

Section B:

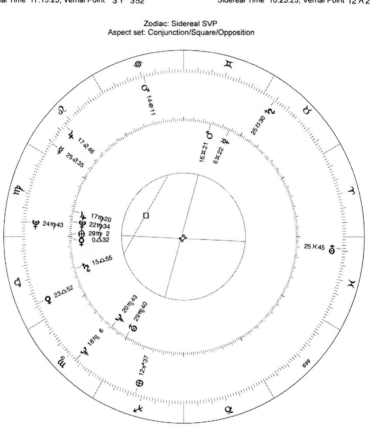

Comparison Chart

Outer - Heliocentric
Birth of John the Baptist
At Jutta (Yattah), Latitude 31N32', Longitude 35E6'
Date: Wednesday, 4/JUN/2 B.C., Julian
Time: 0:37, Local Time
Sidereal Time 17:15:23, Vernal Point 3 ♈ 3'52"

Inner - Heliocentric
Birth of Raphael
At Urbino/Italy, Latitude 43N44', Longitude 12E38'
Date: Friday, 28/MAR/1483, Julian
Time: 21:27, Local Time
Sidereal Time 10:25:25, Vernal Point 12 ♓ 27'56"

Zodiac: Sidereal SVP
Aspect set: Conjunction/Square/Opposition

*Chart 5: heliocentric: the birth horoscope of John the Baptist
with the birth horoscope of Raphael*

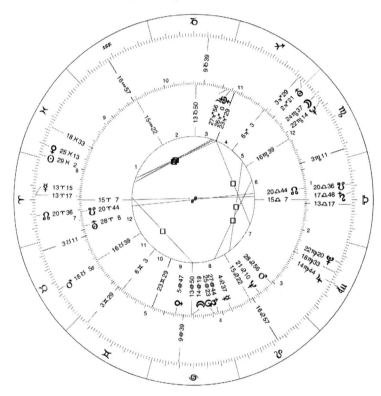

*Chart 6: g: the birth horoscope of Raphael
with the conception horoscope of Novalis*

Note the exact alignment of the Moon's Nodes from one
incarnation to the next.

Comparison Chart

Outer - Tychonic
Birth of Raphael
At Urbino/Italy, Latitude 43N44', Longitude 12E38'
Date: Friday, 28/MAR/1483, Julian
Time: 21:27, Local Time
Sidereal Time 10:25:25, Vernal Point 12ℋ27'56"

Inner - Tychonic
Epoch of Novalis
At Wiederstedt, Latitude 51N38', Longitude 11E30'
Date: Friday, 9/AUG/1771, Gregorian
Time: 21: 7, Time Zone GMT
Sidereal Time 19: 5:17, Vernal Point 8ℋ26'49"

House System: Placidus, Zodiac: Sidereal SVP
Aspect set: Conjunction/Square/Opposition

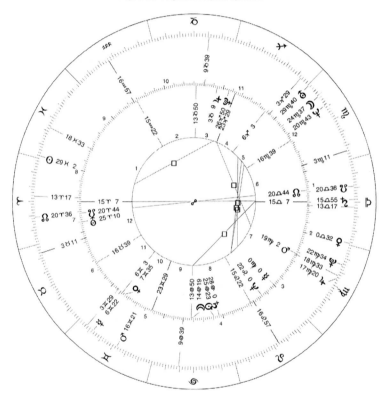

*Chart 7: ♄: the birth horoscope of Raphael
with the conception horoscope of Novalis*

Heliocentric Venus at the conception of Novalis aligned with
heliocentric Mercury at the birth of Raphael, which in turn
aligned with heliocentric Mercury at the Baptism in the Jordan
(see charts 2 & 3).

Comparison Chart

Outer - Geocentric	Inner - Geocentric
Death of Raphael	Epoch of Novalis
At Rome, Italy, Latitude 41N53', Longitude 12E30'	At Wiederstedt, Latitude 51N38', Longitude 11E30'
Date: Friday, 6/APR/1520, Julian	Date: Friday, 9/AUG/1771, Gregorian
Time: 12: 0, Local Time	Time: 21: 7, Time Zone GMT
Sidereal Time 1:36:42, Vernal Point 11 ♓56'59"	Sidereal Time 19: 5:17, Vernal Point 8 ♓26'49"

House System: Placidus, Zodiac: Sidereal SVP
Aspect set: Conjunction/Square/Opposition

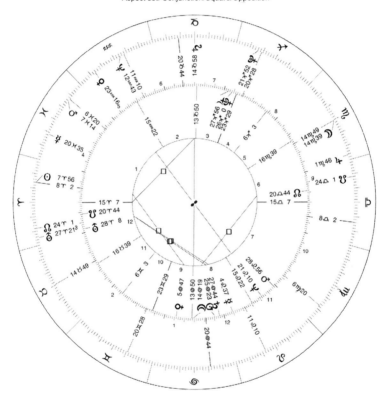

Chart 8: g: the death horoscope of Raphael
with the conception horoscope of Novalis

In addition to the Uranus alignment, the Moon at the conception of Novalis aligned on the opposite side of the zodiac, with Saturn at the death of Raphael.

Comparison Chart

Outer - Geocentric	Inner - Geocentric
Birth of Raphael	Birth of Novalis
At Urbino/Italy, Latitude 43N44', Longitude 12E38'	At Wiederstedt, Germany, Latitude 51N38', Longitude 11E30'
Date: Friday, 28/MAR/1483, Julian	Date: Saturday, 2/MAY/1772, Gregorian
Time: 21:27, Local Time	Time: 10: 8, Local Time
Sidereal Time 10:25:25, Vernal Point 12 ✵ 27'56"	Sidereal Time 0:51: 0, Vernal Point 8 ✵ 26'12"

House System: Placidus, Zodiac: Sidereal SVP
Aspect set: Conjunction/Square/Opposition

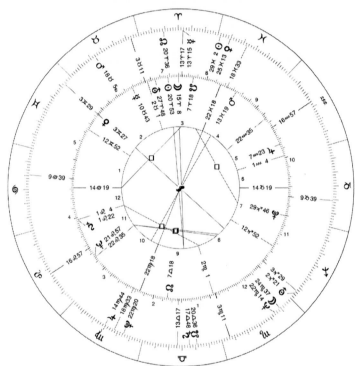

*Chart 9: g: the birth horoscope of Raphael
with the birth horoscope of Novalis*

The Sun at the birth of Novalis aligned exactly with the
Moon's North Node at the birth of Raphael.

Comparison Chart

Outer - Geocentric	Inner - Geocentric
Death of Raphael	Birth of Novalis
At Rome, Italy, Latitude 41N53', Longitude 12E30'	At Wiederstedt, Germany, Latitude 51N38', Longitude 11E30'
Date: Friday, 6/APR/1520, Julian	Date: Saturday, 2/MAY/1772, Gregorian
Time: 12: 0, Local Time	Time: 10: 8, Local Time
Sidereal Time 1:36:42, Vernal Point 11 ♓56'59"	Sidereal Time 0:51: 0, Vernal Point 8 ♓26'12"

House System: Placidus, Zodiac: Sidereal SVP
Aspect set: Conjunction/Square/Opposition

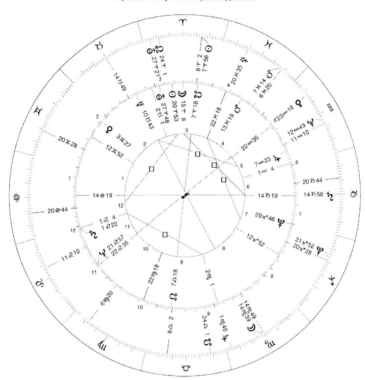

Chart 10: g: the death horoscope of Raphael
with the birth horoscope of Novalis

The planet Uranus at the birth of Novalis aligned exactly with
Uranus at the death of Raphael, having made three orbits
of the zodiac. Also, the Moon's South Node at the birth of
Novalis aligned with the Sun at the death of Raphael.

Comparison Chart

Outer - Tychonic	Inner - Tychonic
Death of Raphael	Birth of Novalis
At Rome, Italy, Latitude 41N53', Longitude 12E30'	At Wiederstedt, Germany, Latitude 51N38', Longitude 11E30'
Date: Friday, 6/APR/1520, Julian	Date: Saturday, 2/MAY/1772, Gregorian
Time: 12: 0, Local Time	Time: 10: 8, Local Time
Sidereal Time 1:36:42, Vernal Point 11 ♓ 56'59"	Sidereal Time 0:51: 0, Vernal Point 8 ♓ 26'12"

House System: Placidus, Zodiac: Sidereal SVP
Aspect set: Conjunction/Square/Opposition

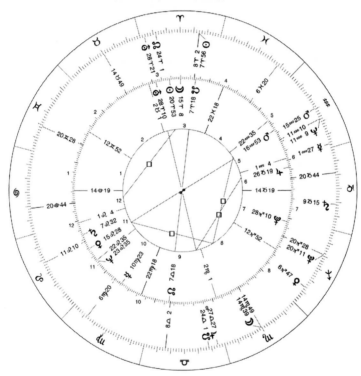

Chart 11: h: the death horoscope of Raphael
with the birth horoscope of Novalis

Horoscopes not included in appendix 2 or in the text of the first five chapters of this book are posted under "Elijah Come Again" on the website www.astrogeographia.org, where all of the horoscopes and comparison horoscopes referred to in this book can be found in color.

NOTES

1   Rudolf Steiner, *The Fall of the Spirits of Darkness* (London: Rudolf Steiner Press, 1993), lecture of October 14, 1917.

2   For those readers unfamiliar with Rudolf Steiner (1861–1925) some excellent biographies are available. There is a summary of his life and activity in *Chronicle of the Living Christ* (see bibliography for all references). In the two-volume work *Hermetic Astrology*, there is an extensive review of Rudolf Steiner's biography from an astrological point of view (volume II) and the results of astrological reincarnation research concerning Rudolf Steiner's previous incarnation (volume I).

3   Rudolf Steiner, *Isis Mary Sophia: Her Mission and Ours*, pp. 213 ff.

4   Rudolf Steiner, *Karmic Relationships*, vol. IV, p. 124.

5   Novalis, *Werke*, p. 98.

6   Florian Roder, *Novalis*, p. 516. Here, with the word "himself" is meant the transcendental, higher Self.

7   Rudolf Steiner, *Karmic Relationships*, vol. VII, p. 62.

8   Robert Powell, *The Christ Mystery: Reflections on the Second Coming.*

9   Valentin Tomberg, *Inner Development*, p. 30.

10 See the bibliography.

11 The dates of birth or death are not always known, in which case it is not possible to cast the horoscope of birth or death of the person—as in the case of the Old Testament prophet Elijah discussed in this book, since it is not known in which year he was born or in which year he died, let alone the actual dates of birth and death.

12 Rudolf Steiner, *Karmic Relationships*, vol. IV, p. 111.

13 Rudolf Steiner, *The Last Address*, p. 18.

14 "Laws" is written in quotes since it is a matter not of hard-and-fast laws but of flexible rules.

15 Rudolf Steiner, *Life between Death and Rebirth*, p. 97.

16 What is meant here by "in the same way or in a metamorphosed way" is that the angle between Saturn and the Sun is either the same, as in the example of Rudolf Steiner considered in this chapter, or is complementary with respect to 180 degrees. For example, if the angle between the Sun and Saturn at death in one incarnation was 50 degrees, it could be 50 degrees again or it might be 130 degrees at birth in the next incarnation, since the sum of the two angles (50+130) is 180 degrees, signifying that the two angles are complementary to one another with respect to 180 degrees.

17 Rudolf Steiner, *Complete Works,* vol. 204, lecture of May 5, 1921. Words in brackets [ ] added by R.P.

18 The horoscopes are arranged alphabetically, and when it is a matter of a comparison horoscope, for example between the birth horoscope of

Rudolf Steiner and the death horoscope of his previous incarnation as Thomas Aquinas, the comparison horoscope is listed according to the later incarnation—in this case: Rudolf Steiner (cataloged as Steiner, Rudolf).

19   Robert Powell, *History of the Zodiac* gives an overview of the ancient sidereal zodiac of the Babylonians, Egyptians, and Ancient Greeks, as well as including a precise definition of the sidereal zodiac, which is still used to the present day in India, albeit in a variant form. The use of the sidereal zodiac as the basis for astrological reincarnation research is elaborated in *Hermetic Astrology*, vol. I, chapters 1 and 2.

20   See Margarete and Erich Kirchner-Bockholt, *Rudolf Steiner's Mission and Ita Wegman,* concerning Rudolf Steiner's previous incarnation as Thomas Aquinas. For readers unfamiliar with the symbols for the astrological signs, see the tabulation at the start of appendix 2.

21   *Hermetic Astrology*, vol. 1, p. 375.

22   Ibid., p. 292.

23   Robert Powell, *History of the Zodiac*, pp. 80–83.

24   A further refinement is the use of the *hermetic horoscope* cast in terms of the sidereal zodiac. The term "hermetic" indicates an extension of the heliocentric horoscope in the sense of Tycho Brahe's astronomical system in which not only the heliocentric positions of the planets are referred to but also the geocentric positions of the Sun, the Moon, the Moon's Node, etc., are included in the hermetic horoscope. See *Hermetic Astrology*, vol. I, for further details concerning the significance of the hermetic system. The Tychonic/hermetic astronomical system includes the heliocentric, Sun-centered system and adapts it to the vantage point of the Earth—thus taking into account the incarnation of the soul coming from the Sun to the Earth.

25   Robert Powell, *The Most Holy Trinosophia: The New Revelation of the Divine Feminine*, pp. 100–109, 132–148.

26   Valentin Tomberg, *Christ and Sophia: Anthroposophical Meditations on the Old Testament, New Testament, and Apocalypse*, pp. 302–313.

27   Novalis: *Pollen and Fragments*, Aphorism 301 (trsl., p. 81).

28   Ibid., Aphorism 52 (trsl., p. 34).

29   Excavations in the late 1990s uncovered a large Byzantine "Church of Elijah" built at this location to commemorate his birthplace. The church, located at Tell Mar Elias near the town of Ajlun in Jordan, stands on the summit of a large hill overlooking the green mountains of Gilead and the rolling hills and plains that lead down toward the Jordan Valley and the Sea of Galilee—http://www.asor.org/outreach/Features/elias.htm.

30   There is no comprehensive biography of Willi Sucher (1902–1985) available at the present time. His books and publications are available from the Astrosophy Research Center, P.O. Box 13, Meadow Vista, California, 95722. See also Willi Sucher, *Cosmic Christianity*

& the Changing Countenance of Cosmology, pp. 9–24: Biographical Introduction.

31   Christian Hermetic Astrology; and Chronicle of the Living Christ.

32   This was not the officially defined astronomical zodiac of the International Astronomical Union, but an approximation to it drawn up initially by Elisabeth Vreede in 1929 and later slightly altered by Suso Vetter in 1972. This version of the astronomical zodiac is used to the present day in the yearly Sternkalender (Star Calendar) published by the Mathematical-Astronomical Section at the Goetheanum in Dornach, Switzerland, founded by Rudolf Steiner. The astronomical zodiac comprises twelve unequal constellations.

33   The sidereal zodiac, which originated with the Babylonians, is defined in relation to the fixed stars. See Robert Powell, History of the Zodiac, appendix 1 for a formal definition of the Babylonian sidereal zodiac.

34   Robert Powell, Chronicle of the Living Christ, Appendix V.

35   The birth of John the Baptist was six months before the birth of Jesus of Nazareth, which was shortly before midnight on December 6, 2 B.C.E. See Chronicle of the Living Christ, p. 149.

36   See Chronicle of the Living Christ for the conception, birth, and death horoscopes of Jesus and John the Baptist.

37   Robert Powell, History of the Zodiac, chapter 3 provides a more detailed description of the background of the tropical zodiac, and how it came to be used by Arab astrologers, replacing the sidereal zodiac of the Babylonian, Egyptian, and Greek astrologers.

38   Historians use the system 3 B.C.E., 2 B.C.E., 1 B.C.E., C.E. 1, C.E. 2, etc. going straight from 1 B.C.E. to C.E. 1, without the year "zero." Astronomers, in order to include a year "zero," use the system -2, -1, 0, +1, +2, etc., simply equating 1 B.C.E. with the year "zero" and using minus dates going back from 1 B.C.E.. The astronomical system enables one to compute intervals between dates accurately and precisely.

39   The term "hermetic" signifies a heliocentric perspective (the perspective of the Sun), which includes the location of the Sun and the Moon as well as the Ascendant, Descendant, and Midheaven points, none of which are apparent from the heliocentric perspective. The hermetic perspective was applied in ancient Egypt and was formulated astronomically by the Danish astronomer Tycho Brahe. Thus the hermetic perspective is also referred to as the Tychonic system. See Hermetic Astrology, vol. I, chapter 2.

40   The ancient Egyptian hermetic perspective is almost identical to that of the Tychonic astronomical system put forward by the Danish astronomer Tycho Brahe (1546–1601). In his system the planets revolve around the Sun, but the Sun, Moon, and Moon Nodes revolve around the Earth—the Earth being central. Because of this, it is possible to compute also the Ascendant and Descendant, the Midheaven and other house cusps, and include them in the hermetic/Tychonic horoscope, which is thus a helio-geocentric horoscope. Essentially it is a geocentric

horoscope in which the planetary positions are heliocentric (except for the positions of the Sun and Moon, which are geocentric).

41 Kepler held to the Copernican (heliocentric) astronomical system that led to his formulation of the three laws of planetary motion. See *Hermetic Astrology*, vol. I, chapter 2.

42 See *Hermetic Astrology*, vol. I, chapter 4 for an account of these two levels.

43 In the practice of astrology, the chart is referred to simply as the hermetic horoscope—implicitly the heliocentric planetary positions are included in the hermetic chart.

44 See end of bibliography for details. The Astrofire research program developed by Peter Treadgold (1943–2005) contains a research module, star catalogue, and database of birth and death dates of historical personalities. In the Astrofire program the hermetic horoscope is referred to as the Tychonic chart in honor of Tycho Brahe.

45 As discussed in *Hermetic Astrology*, vol. I, chapter 10, the epoch horoscope found by applying the hermetic rule indicates the "etheric conception" that coincides fairly closely in time with the physical conception. The epoch is the moment when the soul begins building up its etheric/life body.

46 The hermetic rule is an ancient astrological rule for determining the horoscope of conception retrogressively from the day and hour of birth. It is named after Hermes, the great teacher of the Egyptian mystery wisdom, who is said to be the originator of this astrological rule linking conception and birth. Hermetic astrology is so named because it works extensively with both conception and birth, and thus is based upon the use of the "rule of Hermes." See *Hermetic Astrology*, vol. I, appendix 1 for the history of the hermetic rule.

47 During the hours of sleep the "I" indwells the astral body and is released from the physical and etheric bodies while continuing to maintain a connection with them.

48 From the English burial service adapted from the text of Genesis 3:19.

49 In the case of the epoch, the Astrofire program works with GMT = Greenwich Mean Time, whereby 9:53 P.M. local time at Wiederstedt corresponds to 9:07 P.M. at London/Greenwich.

50 The Moon is referred to as the "gateway to birth," and in this case it aligns with the "gateway to death."

51 For the Babylonians and Egyptians, when Jupiter was seen against the background of the star cluster Praesepe, this was considered to be the most powerful location of Jupiter in the whole zodiac and was referred to as Jupiter's place of exaltation.

52 Each sign of the zodiac is divided into three 10° sections known as *decans*. This alignment is in the third decan of Cancer, which extends above the region of the flank of the Great Bear, the constellation of Ursa Major. See Lacquanna Paul & Robert Powell, *Cosmic Dances of the Zodiac* concerning this decan and the decans in general.

53   Robert Powell, *Chronicle of the Living Christ*, p. 152 gives the birth horoscope of Mary of Nazareth (Nathan Mary).

54   *Sacred Song XIV*, translated by Eileen Hutchins.

55   *Sacred Song XV*, translated by Eileen Hutchins.

56   *Chronicle of the Living Christ*, p. 161.

57   See, for example, Niek Scheps, *The Trutine of Hermes: A Guide to Calculating and Interpreting the True Ascendant.*

58   "Christendom or Europe," *Hymns to the Night and Other Selected Writings*, p. 58.

59   Michel Gauquelin, *The Spheres of Destiny: Your Personality and the Planets* presents empirical research demonstrating the great significance of the planets close to the Ascendant or Midheaven, a relationship that shows itself—in a statistically significant way—in the person's personality traits and profession.

60   Lacquanna Paul & Robert Powell, *Cosmic Dances of the Planets*, pp. 76–90 & 182–186.

61   Robert Powell, *The Sign of the Son of Man in the Heavens*, pp. 119–120 clarifies the relationship of Christ (and his teaching) to the sign of Pisces.

62   J. E. Zeylmans van Emmichoven, *A New Novalis?* p. 12.

63   Heinz Ritter-Schaumburg, *Novalis und seine erste Braut*, figure 1.

64   Frederick Hiebel, *Novalis*, p. 64.

65   *Sacred Songs of Novalis*, p. 23.

66   *Hymns to the Night and Other Selected Writings*, p. 5.

67   Frederick Hiebel, *Novalis,* p. 27.

68   Ibid., p. 70.

69   *Sacred Songs of Novalis*, pp. 15–17.

70   Ibid., p. 21.

71   "Christendom or Europe," *Hymns to the Night and Other Selected Writings,* p. 60.

72   Frederick Hiebel, *Novalis*, p. 33.

73   Ibid., p. 31.

74   Ibid., p. 36.

75   Nello Ponente, *Who was Raphael?*, pp. 13–14.

76   *Grosse Männer der Weltgeschichte* (Great Men in World History), p.378.

77   Robert Powell, *Chronicle of the Living Christ*, p. 152 gives the birth horoscope of the Nathan Mary (Mary of Nazareth).

78   Vasari, quoted by Nello Ponente, *Who is Raphael?*, p. 75.

79   Robert Powell, *The Christ Mystery: Reflections on the Second Coming* explores the dating of this great mystery of our time and comes to the conclusion that the onset of the second coming began in 1933 and is continuing in our time now, in the twenty-first century.

80   Robert Powell, *Hermetic Astrology*, vol. I, appendix 2, pp. 330–356 discusses the first "law" in detail.

81   See *Hermetic Astrology*, vol. 1, p. 340, where the example of Caliph
      Mu'awiya (p. 440) is compared with a later incarnation as President
      Woodrow Wilson (p. 454). In this reincarnation example there is also
      an alignment of Uranus at the birth of Woodrow Wilson with the posi-
      tion of Uranus at the death of Mu'awiya. Whereas between the death of
      Raphael and the birth of Novalis Uranus orbited the zodiac three times,
      between the death of Mu'awiya and the birth of Woodrow Wilson
      Uranus orbited the zodiac fourteen times.

82   Bearing in mind that the heliocentric-Mercury position in sidereal
      Aquarius in Raphael's death horoscope is an average one, h-Mercury
      might have been two degrees more or less advanced.

83   Frederick Hiebel, *Novalis*, p. 52: "When a spirit dies, it becomes a
      human being. When a human being dies, it becomes a spirit."—from
      the *Fragments* of Novalis.

84   Walter Johannes Stein, "Ein Beitrag zu Rudolf Steiners Lebensgeschichte
      (nach einem Gespräch mit Rudolf Steiner)," *Korrespondenz der anthro-
      posophischen Arbeitsgemeinschaft*, vol. 3 (February, 1934), pp. 17–21.
      English translation: Walter Johannes Stein, "A Contribution to the
      History of Rudolf Steiner's Life," *Anthroposophical Quarterly*, vol. 17
      (1972), pp. 78–81. Cf. also Charles Lawrie, "For the Conscience of
      Europe," *Golden Blade* (1991), p. 68. It has to be noted that although
      Rudolf Steiner indicated that he had been unable to carry out his own
      destiny (elaborating upon the concrete knowledge of reincarnation and
      karma) in 1924 he did then make a start with his true mission in giving
      the series of 84 lectures entitled *Karmic Relationships*.

85   There are nine possible chart comparisons of epoch, birth, death with
      epoch, birth, death in two incarnations. However, taking account of
      the two levels—geocentric and heliocentric/hermetic—there are eigh-
      teen chart comparisons.

86   See *Hermetic Astrology*, vol. 1, pp. 388–393.

87   Robert Powell, *Hermetic Astrology*, vol. 1, pp. 171–172—translated by
      R.P. from the original German report of Max Gümbel-Seiling published
      in: *Gedenkblatt für Marie Steiner geb. von Sievers* (Stuttgart, 1949), p.
      13. According to Max Gümbel-Seiling, this was the turn of the year
      1908/1909, and Rudolf Steiner then spoke of the mystery of Elijah's
      incarnations at a celebration in Munich on January 6, 1909.

88   Erica von Dellingshausen, *Die Wartburg*, p. 55.

89   Rudolf Steiner, *Complete Works*, vol. 109, lecture of April 6, 1909.
      This lecture was published in English translation under the title "Avatar
      Beings" in: Rudolf Steiner, *Christianity in Human Evolution*.

90   Erica von Dellingshausen, *Die Wartburg*, pp. 65–66.

91   Erica von Dellingshausen, *Die Wartburg*, see p. 24 for historical dat-
      ing of the first and second Battle of the Minstrels. (In regard to Rudolf
      Steiner's indication: Rudolf Steiner places the Battle of the Minstrels in
      1206 and refers to a second Battle of the Minstrels but without dating
      it to the year 1207.)

92 Rudolf Steiner, *Karmic Relationships,* vol. 4, p. 108.

93 Erica von Dellingshausen, *Die Wartburg,* p. 36.

94 Rudolf Steiner, *The Last Address,* Introduction, p. 10: "At the awakening of Lazarus, the spiritual being John the Baptist, who since his death had been the overlighting spirit of the disciples, penetrated from above into Lazarus..."

95 Rudolf Steiner, *From the History and Contents of the First Section of the Esoteric School 1904–1914,* p. 226: "The writer of Saint John's Gospel is Lazarus."

96 Frederick Hiebel, *Novalis,* pp. 1–2.

97 Rudolf Steiner, *The Four Seasons and the Archangels,* particularly lectures 3 and 5 (October 7 and October 13, 1923) are about the Archangel Raphael.

98 Virginia Brett, *Healing Sequences from Raphael and Novalis* presents a juxtaposition of the fifteen *Spiritual Songs* of Novalis with the fifteen *Madonna paintings* by Raphael and others in a healing sequence suggested by Rudolf Steiner. This juxtaposition was made by Virginia Brett, who also presents in this work her own translation from the German of Novalis's *Spiritual Songs.*

99 *The Life and Times of Raphael,* p. 60.

100 *The Life and Times of Raphael,* p. 12.

101 Nello Ponente, *Who Was Raphael?,* p. 128.

102 Frederick Hiebel, *Novalis,* p. 5.

103 Adalbert von Keyserlingk, *Monte Gargano,* p.250.

104 A moral code of ethics set forth in *The Book of the Courtier* by Baldassare Castiglione.

105 Frederick Hiebel, *Novalis,* p. 8.

106 J. E. Zeylmans van Emmichoven, *A New Novalis?,* pp. 12, 13, 24.

107 Frederick Hiebel, *Novalis,* p. 34.

108 Adelbert von Keyserlingk, *Monte Gargano,* p. 249, footnote 82. It should be noted that St. Elisabeth's hospital in Marburg is not in Thüringia but, nevertheless, St. Elisabeth had grown up in Thüringia and came from there to Marburg to found the hospital.

109 Frederick Hiebel, *Novalis,* pp. 35, 99.

110 Frederick Hiebel, *Novalis,* p. 43.

111 Novalis, *Henry of Ofterdingen,* p. 148.

112 Novalis, *Henry of Ofterdingen,* pp. 69–72.

113 *The Sacred Songs of Novalis,* trsl. by Eileen Hutchins, Introduction, p. 6.

114 Novalis, "Christendom of Europe" in *Hymns to the Night and Other Selected Writings,* pp. 45–47, 62.

115 Frederick Hiebel, *Novalis,* p. 43.

116 Lacquanna Paul & Robert Powell, *Cosmic Dances of the Planets,* Appendix V, p. 193, discusses the significance of the Galactic Center and its location at 2° Sagittarius in the sidereal zodiac. See also Robert

Powell & Kevin Dann, *Christ & the Maya Calendar: 2012 and the Coming of the Antichrist,* "Appendix I: The Central Sun."

117 Another exception here, in addition to the Moon, is heliocentric Mercury, which moves about 2°45' a day when it is close to aphelion (farthest distance from the Sun) and about 6°20' a day when it is near perihelion (closest distance to the Sun). Taking the noon position of h-Mercury on its perihelion day could therefore lead to an error in its actual location of over three degrees (3°10'), if the event under consideration took place around midnight rather than midday. Although not as large as the possible deviation (around 7°30') in the Moon's position between noon and midnight, this possible deviation in the position of h-Mercury is larger than the tolerable orb of three degrees. Otherwise, the largest possible deviation between the midnight and noon positions of any planet is just under one degree in the case of h-Venus and is much less with all the other planets, including g-Mercury.

118 It is reported that she died at the "first cock crow" on Monday, November 17, 1231, which must have been around 5 AM that day. "In the morning twilight she lay there filled with heavenly joy and visibly in ecstasy until the first cock crow, when she said: 'Behold the hour is approaching when the Virgin gave birth.' In deep devotion she then commended all present to God and departed from life as if in the most sweet sleep." Erica von Dellingshausen, *Die Wartburg,* p. 66.

119 The question is often raised why the Moon is not shown in a heliocentric chart. Since this chart depicts the perspective from the Sun, it is the Earth—as the third planet from the Sun after Mercury and Venus—that is seen. From the Sun's perspective the Moon is simply an appendage to the Earth, rather like an ear attached to a head.

120 In the practice of Astrosophy (Star Wisdom) it is well-established that the heliocentric chart is an indispensable adjunct to the traditional geocentric chart. If the hermetic chart is used *instead of* the heliocentric chart, it contains all the information of the heliocentric perspective plus the additional information derived from the inclusion of the geocentric elements providing a framework for the heliocentric planetary positions. In other words, the hermetic chart is not just equivalent to the heliocentric chart, but goes beyond it and supplements it with important cosmic information that can *only* be derived through the hermetic perspective, as may be seen from the examples under consideration here. Nevertheless, many people accustomed to working with heliocentric charts may prefer to continue using the heliocentric chart alongside the geocentric and hermetic charts. (Note that hermetic = Tychonic, this being the helio-geocentric astronomical system introduced by the Danish astronomer Tycho Brahe.)

121 Rudolf Steiner, *Karmic Relationships,* vol. V, p. 99.

122 Lacquanna Paul & Robert Powell, *Cosmic Dances of the Planets,* appendix 3 discusses the chakras in relation to the planets.

123 Rudolf Steiner, "The Spiritual Individualities of the Planets," *Golden Blade* (1988), pp. 44–47.

124 Lacquanna Paul & Robert Powell, *Cosmic Dances of the Planets*, appendix 3 discusses the chakras in relation to the planets. See also Robert Powell, *Hermetic Astrology*, vol. I, chapter 5.

125 Robert Powell, *Hermetic Astrology*, vol. I, pp. 135–137 discusses the symbolism of the six-pointed star, or "Seal of Solomon," in relation to the six-petal lotus flower.

126 Rudolf Steiner, *Esoteric Christianity and the Mission of Christian Rosenkreutz*, lecture of November 5, 1911, indicates the bodhisattva status of Abraham, Moses, and Elijah.

127 Rudolf Steiner means, of course, the secret doctrine of the Hebrew people, and his remark is confirmed by the following statement: "The Midrash and the Zohar say that Phinehas has the same soul as Elijah the prophet, who will usher in the Messiah"—Rabbi Tuvia Bolton, "Phinehas: Justified by his Faith," p. 9.

128 Rudolf Steiner, *Gospel of St. Mark*, p. 146, words in [ ] added by R.P.

129 This refers to the form of Baal worship as it was encountered by the people of Israel in the way it was practiced at Peor, the mountain peak to which Balak led Balaam to induce him to pronounce a curse upon Israel. Baal was a Canaanite fertility deity whose ceremonies often included human sacrifice and temple prostitution. Baal-Peor means simply "Baal of Peor," and signifies the worship of Baal practiced there by the Moabites under King Balaak. Cf. Valentin Tomberg, "Yahweh and Baal in the Destiny of Israel," *Christ and Sophia*, pp. 86–88.

130 Rabbi Tuvia Bolton, "Phinehas: Justified by his Faith."

131 *The Zohar*, trsl. Maurice Simon and Harry Sperling, vol. V, p. 323.

132 Rudolf Steiner, *The Last Address*, Introduction, p. 10.

133 Charles Tidball & Robert Powell, *Jesus, Lazarus, and the Messiah*, pp. 96–109 discusses this "John mystery." See also Robert Powell, *The Mystery, Biography, and Destiny of Mary Magdalene*, chapter 3.

134 *The Zohar*, trsl. Maurice Simon and Harry Sperling, vol. V, p. 325. Nadab and Abihu were the two eldest sons of Aaron, who were appointed by God to the priesthood, but who were struck down by fire from the Lord because they "offered profane fire before the Lord which he had not commanded them" (Leviticus 10:1). Aaron was the first high priest of Israel (together with Moses). Through the death of his two eldest sons, his third son, Eleazor, became the next high priest, and Phinehas, the son of Eleazor, became the third high priest of Israel.

135 Ibid., p. 330.

136 Ibid., p. 227.

137 Ibid., p. 324.

138 Ibid., p. 323.

139 Ibid., p. 324.

140 Ibid., p. 335.

141 Rudolf Steiner, *According to Luke*, p. 122.

142 Ibid., p. 128.

143 Ibid., p. 121.

144 Ibid., p. 128.

145 Rudolf Steiner, *Esoteric Christianity and the Mission of Christian Rosenkreutz*, lecture of November 5, 1911, indicates the bodhisattva status of Abraham, Moses, and Elijah.

146 Rudolf Steiner, *According to Luke*, p. 95. Rudolf Steiner used the expression "Nathan Jesus" to signify the Jesus child whose birth is described in the *Gospel of Luke*. He spoke of the instreaming grace of the Buddha making itself felt especially in the proclamation to the shepherds: "Peace on Earth to all human beings of good will" (Luke 2:14).

147 Ibid., p. 129.

148 Ibid., p. 131.

149 Ibid., p. 130. The intimate connection of the Elijah/John the Baptist individuality to Gautama Buddha can be grasped against the background of an understanding of Elijah as a bodhisattva, thus connected spiritually with Gautama, who also belonged to the circle of bodhisattvas before he attained buddhahood (and, of course, as a Buddha he continued to remain spiritually united with the other bodhisattvas).

150 Ibid., p. 133.

151 Ibid., p. 132.

152 Ibid., p. 112.

153 Robert Powell, *Chronicle of the Living Christ*, chapters on the Nathan Mary, the Nathan Jesus, and John the Baptist give an account of the dates of conception and birth of these individuals, the date of the visitation of the Nathan Mary to Elizabeth, etc.

154 Rudolf Steiner, *According to Luke*, p. 122.

155 Rudolf Steiner, *The Apocalypse of St. John*, p. 161.

156 Ibid., p. 148.

157 Anne Catherine Emmerich, *The Life of Jesus Christ*, vol. I, pp. 317–318.

158 Ibid., pp. 318–319.

159 Ibid., p. 320.

160 Ibid., pp. 320–321.

161 Ibid., p. 318. These words were spoken by the Angels to the seer, Anne Catherine Emmerich, upon seeing the young six-year-old boy abandoned in the desert by his mother, Elizabeth, in her attempt to hide him from Herod.

162 Ibid., pp. 322–323.

163 Anne Catherine Emmerich, *Lehrjahre Jesu*, vol. 1, pp. 36–37.

164 Rudolf Steiner, *The Apocalypse of St. John*, p. 161.

165 Rudolf Steiner, *Gospel of St. Mark*, p. 119.

166 Anne Catherine Emmerich, *The Life of Jesus Christ*, vol. I, p. 425.

167 Rudolf Steiner, *The Last Address*, p. 10.

168 Ibid.

169 Valentin Tomberg, *Christ and Sophia*, p. 326.

170 On account of the hermetic rule, the birth Ascendant is the same as the location of the Moon at epoch/conception, which is also at 14½° Cancer.

171 This finding is in accordance with Rudolf Steiner's description in his lecture "Hidden Sides of Man's Existence and the Christ Impulse," *Anthroposophical Quarterly*, vol. 17 (1972), pp. 42–51: "We actually experience the constellations of the fixed stars, those constellations that were known in ancient times as the zodiac...experiencing one's own being as an image of the heaven of fixed stars...midway between death and a new birth...From that other world, while we are descending [back into incarnation for a new life on Earth], we see...the fixed stars appear and behind them the planetary movements." Words in [ ] added by R.P.

172 Novalis, *Werke*, p. 436.

173 In hermetic astrology the sidereal zodiac is recognized to be an outer expression of the Akasha Chronicle. Hence this statement by Rudolf Steiner indicates the whole basis of ancient astrology, which was to look at the movements of the planets against the background of the stars of the sidereal zodiac.

174 Rudolf Steiner, *Life between Death and Rebirth*, pp. 44, 141–142.

175 See the quote by Novalis at the start of this chapter for the precise formulation.

176 Rudolf Steiner, *The Last Address*, p. 10: "At the awakening of Lazarus, the spiritual being of John the Baptist, who since his death had been the overshadowing Spirit of the disciples, penetrated from above into Lazarus as far as the consciousness soul. The being of Lazarus himself, from below, intermingled with the spiritual being of John the Baptist from above. After the awakening of Lazarus, this being is Lazarus–John, the disciple whom the Lord loved."

177 Ibid., pp. 12–15.

178 Rudolf Steiner, *Karmic Relationships*, vol. VI, p. 110.

179 Ibid.

180 Ibid.—Rudolf Steiner uses the Greek word "Archai," whom he refers to as Time Spirits, whereas Dante uses the Latin word "Principalities" to describe the celestial hierarchy indwelling the Venus sphere.

181 Frederick Hiebel, *Novalis*, pp. 18–19.

182 Ibid., p. 21.

183 Ibid., p. 25.

184 Rudolf Steiner, lecture of July 27, 1923, in *Collected Works*, vol. 228, pp. 10–11 (translated from the German by R.P.). An English translation of this lecture was published under the title "The Spiritual Individualities of the Planets," in the *Golden Blade* (1988), pp. 43–49. In the most recent German edition of this lecture, the wording is reversed: "If Venus stands square to Mars, a language becomes soulless, shrill... if Venus and Mars stand in conjunction... a language becomes inwardly deepened, full of soul."

185 Frederick Hiebel, *Novalis*, p. 46.

186   Rudolf Steiner, *Man: Hieroglyph of the Universe*, pp. 56–57: "...[there] opens a window facing quite another world...our world is at these times penetrated by astral streams...it breathes in the astral world."

187   Rudolf Steiner, *Karmic Relationships*, vol. V, pp. 92–93. Rudolf Steiner used the Greek names *Exusiai, Dynamis,* and *Kyriotetes* to describe the three ranks of celestial hierarchies indwelling the Sun.

188   See Rudolf Steiner, *From the History & Contents of the First Section of the Esoteric School 1904–1914*, p. 225 concerning Rudolf Steiner's indication of the Master Jesus as the Friend of God. Dionysius the Areopagite, *Mystical Theology and the Celestial Hierarchies*, p. 48, also refers to Melchizedik as "the Friend of God."

189   Rudolf Steiner, *The Spiritual Hierarchies*, pp. 98–99, 126—words in [ ] added by R.P.

190   Rudolf Steiner, *Karmic Relationships*, vol. 5, p. 97.

191   Ibid., pp. 85, 88–89.

192   Rudolf Steiner, *True and False Paths in Spiritual Investigation*, p. 131—words inserted in [ ] by R.P.

193   Rudolf Steiner, *The Spiritual Hierarchies*, p. 126.

194   In 2006 the International Astronomical Union demoted Pluto from its status as a planet to that of a "dwarf planet." Notwithstanding this demotion, astrological reincarnation research shows conclusively that Pluto is a planet as far as the reincarnating soul is concerned. Consider, for example, the fact referred to in chapter 1 that Raphael reincarnated after Pluto had made exactly six orbits of the sidereal zodiac from the time of birth of John the Baptist. In other words, the soul of Raphael waited to reincarnate until Pluto reached the same location in the sidereal zodiac as at birth in his incarnation as John the Baptist.

195   Robert Powell, *Christian Hermetic Astrology*, pp. 15–25.

196   Rudolf Steiner, *Man's Being, His Destiny, and World Evolution*, pp. 16, 28–29.

197   Rudolf Steiner, *The Spiritual Hierarchies*, pp. 113–114.

198   Rudolf Steiner, *Karmic Relationships*, vol. 6, pp. 114–115.

199   Goethe, *Wilhelm Meister's Years of Travel*, book I, chapter 10.

200   Rudolf Steiner, *Man's Being, His Destiny, and World Evolution*, p. 32.

201   Robert Powell, *Hermetic Astrology*, vol. I, appendix 1, describes the hermetic rule and its historical background.

202   According to the hermetic rule, the Moon at epoch is the same as the Ascendant at birth, which in the case of Rudolf Steiner was 19½° Libra.

203   Ptolemy, *Tetrabiblos* III. 1 (trsl. pp. 223–225).

204   See details of the Astrofire program at the end of the bibliography.

205   Rudolf Steiner, *Man's Being, His Destiny, and World Evolution*, p. 32.

206   See Robert Powell, *The Most Holy Trinosophia and the New Revelation of the Divine Feminine* regarding the current activity of this individuality, referred to in *The Most Holy Trinosophia* as the "third teacher."

207   Frederick Hiebel, *Novalis*, p.50.

208   Ibid., pp. 45, 46, 47.

209 Ibid., *Novalis*, p.23.
210 Novalis, *Pollen And Fragments*, trsl. Arther Versluis, p. 25.
211 Frederich Hiebel, *Novalis*, p. 44.
212 Some authorities identify the two witnesses as Enoch and Elijah.
213 *The Zohar*, vol. V, trsl. by Maurice Simon and Harry Sperling, p. 363.
214 Rudolf Steiner, *Last Address*, pp. 17–18.
215 Novalis, *Hymns to the Night*, no. 5, translated by R.P. from *Werke*, p. 51.
216 Rudolf Steiner, *The Last Address*.
217 Robert Powell, *The Most Holy Trinosophia*, pp. 19–27, 132–148 for a discussion of the Rose of the World as the coming Sophia culture of the new Aquarian age and also concerning the third teacher. Cf. also Daniel Andreev, *The Rose of the World*.
218 *Meditations on the Tarot*, chapter 11, explores the attainment of strength through purity.
219 See Robert Powell, *The Christ Mystery* and also the articles in recent issues of the yearly *Christian Star Calendar* for insight into the coming of the Antichrist and his Prophet as an event in our time. See also Robert Powell & Kevin Dann, *Christ & the Maya Calendar: 2012 and the Coming of the Antichrist*.
220 Rudolf Steiner, *The Last Address*, p. 17.
221 Caitlin Matthews, *Sophia: Goddess of Wisdom*, p. 341.
222 John the Baptist was directly related to Mary of Nazareth, also known as the Nathan Mary because of her line of descent from David via Nathan. The offspring of an Essene family, John the Baptist and the Nathan Mary were second cousins, since Mary's mother Anne and John's mother Elizabeth were first cousins, according to the description of Anne Catherine Emmerich, *The Life of Jesus Christ*, vol. I, pp. 120–124.
223 Robert Powell, *The Sophia Teachings* and *The Most Holy Trinosophia* draw attention to the coming of Divine Sophia, ushering in the Rose of the World, as a work of *collaboration* between Christ and Sophia. In other words, the new Age of Sophia is at the same time a new Age of Christ, referred to—when the focus is upon Christ—as the Age of the Second Coming, this being the true significance of the term "New Age." See also Robert Powell & Kevin Dann, *Christ & the Maya Calendar: 2012 and the Coming of the Antichrist*, chapter 9, "The Rose of the World"; see in particular page 156 for a reproduction of a Sophia icon from the seventeenth century entitled "Sophia, the All-Wisdom of God."
224 J. E. Zeylmans van Emmichoven, *A New Novalis*, p. 25.
225 Thomas Schipflinger, *Sophia-Maria*, pp. 291–296.
226 Anne Catherine Emmerich, *The Life of Jesus Christ*, vol. I, p. 146.
227 Rudolf Steiner, *Concerning the History and Content of the Higher Degrees of the Esoteric School 1904–1914*, p. 468. Available as *"Freemasonry" and Ritual Work* (CW 265).
228 Ibid., p. 471.

229 Rudolf Steiner's statement could also be understood as the *uniting of two streams*, one leading to spiritual heights and one into the depths of the Earth, and that the unity of these two is the "new stream," as outlined in the conclusion of the next-but-one footnote 231.

230 Ibid., p. 455.

231 The quotations in this paragraph are from a (not yet published) document entitled "The Countess Keyserlingk to Mabel Cotterell." Mabel Cotterell translated works of Rudolf Steiner into English and was present when Rudolf Steiner spoke at Penmaenmawr, Wales, in 1923. Regarding Johanna von Keyserlingk, some of her esoteric conversations with Rudolf Steiner were published in Adalbert von Keyserlingk's book *The Birth of a New Agriculture: Koberwitz 1924*. Something of the contents of this unpublished document evidently underlies Adalbert von Keyserlingk's book *Monte Gargano: Europas ältestes Michaelsheiligtum* ("Monte Gargano: Europe's Oldest St. Michael Sanctuary"). Adalbert was the son of Johanna von Keyserlingk. On page 248 of his book *Monte Gargano* he describes the connection between Buddhism and Christianity through the *bodhisattva stream*. And it may be added that John the Baptist, as a bodhisattva, belonged to this stream. Also, Rudolf Steiner spoke of "John the Baptist's preaching as a revival of the Buddha's sermon" (*Gospel According to Luke*, p. 131). On page 160 of *The Birth of a New Agriculture* Keyserlingk briefly mentions the path of Lazarus–John into the depths of the Earth. This path is elaborated in detail in his book *Monte Gargano* where, on page 298, he refers to the goal of uniting the two paths through initiation, "so that the pure fire of the Earth-Sun-Soul arises within and the spiritual rays of cosmic light stream in from above."

232 Rudolf Steiner, *The Last Address*, p. 17–18, word in [ ] added by R.P.

233 Novalis, *Sacred Songs* XII, trsl. Eileen Hutchins, p. 47.

234 Rudolf Steiner, *The Last Address*, pp. 17–18 (italics added by R.P.).

235 Robert Powell, *The Most Holy Trinosophia*, pp. 66, 141. The significance of the third teacher—in contrast to the first two teachers—appearing in female form can be seen as a metamorphosis from two thousand years ago, when Christ incarnated physically into a male body and had twelve male disciples. Now his incarnation is in an etheric body, and on account of the polarity between the physical and etheric bodies, it is understandable that Christ's manifestation in an etheric body is being proclaimed by a woman. (The polarity referred to here is that the etheric body of a male person is female, and the etheric body of a female person is male.)

236 Ibid. p. 69.

237 Ibid. p. 104; "around the year 1975" (this date is approximate).

238 Robert Powell, *The Most Holy Trinosophia*, p. 141.

239 Ibid., p. 104. The date 1975 is approximate.

240 Robert Powell & Kevin Dann, *Christ & the Maya Calendar: 2012 and the Coming of the Antichrist*.

241 Robert Powell, *The Most Holy Trinosophia*, p. 141.

242 Ibid, p. 153: "The three teachers referred to here are, so to speak, the three 'elders' of the metamorphosed community of the Etheric Christ in the twentieth century. *However, there are more teachers than these three we are considering.*"

243 Rudolf Steiner, *The Last Address*, pp. 17–18.

244 Rudolf Steiner, *Esoteric Christianity and the Mission of Christian Rosenkreutz* (CW 130), lecture of February 9, 1912.

245 Anonymous, *Meditations on the Tarot*, pp. 361–362.

246 This statement has to be qualified in terms of the degree of accuracy with which the time is known.

BIBLIOGRAPHY

Andreev, Daniel, *The Rose of the World* (trsl. Jordan Roberts; Great Barrington, MA: Lindisfarne Books, 1997).

Anonymous, *Meditations on the Tarot: A Journey into Christian Hermeticism* (trsl. Robert Powell; New York: Tarcher/Putnam, 2002).

Bolton, Rabbi Tuvia, *Phinehas: Justified by his Faith;* www.ldolphin.org/Phine.html (URL is case-sensitive).

Bortolon, Liana, *The Life and Times of Raphael* (trsl. B. Paterson; Feltham, England: Paul Hamlyn, 1968).

Brett, Virginia, *Healing Sequences from Raphael and Novalis* (Waianae, HI: Sophia Press, 1991).

Castiglione, Baldassare, *Il Libro del Cortegiano* (Rome, 1528)—(trsl. Sir Thomas Hoby, *The Book of the Courtier;* London, 1561; republished London: David Nutt, 1990).

Dionysius the Areopagite, *Mystical Theology and the Celestial Hierarchies* (Fintry, England: The Shrine of Wisdom, 1965).

Emmerich, Anna Katharina, *Lehrjahre Jesu* (Clemens Brentano *Complete Works*, vols. 24.1 & 24.2; Stuttgart, Germany: Kohlhammer, 1983 & 1985).

———, *The Life of Jesus Christ* (vols. I–IV; Rockport, IL: Tan Books, 2004).

Gauquelin, Michel, *The Spheres of Destiny: Your Personality and the Planets* (London: J.M. Dent & Sons, 1980).

Goethe, *Wilhelm Meister's Years of Travel* (trsl. H.M. Waidson; Richmond, Surrey, United Kingdom: Calder Publications, 1980).

Grosse Männer der Weltgeschichte ("Great Men of World History") (Klagenfurt, Germany: Neuer Kaiser Verlag, 1987).

Hiebel, Frederick, *Novalis* (Chapel Hill, NC: University of North Carolina Press, 1959).

Jocelyn, Beredene, *Citizens of the Cosmos: Life's Unfolding from Conception through Death to Rebirth* (Great Barrington, MA: SteinerBooks, 2009).

Jocelyn, John, *Meditations on the Signs of the Zodiac* (Great Barrington, MA: SteinerBooks, 2006).

Kirchner-Bockholt, Margarete & Erich, *Rudolf Steiner's Mission and Ita Wegman* (London: Rudolf Steiner Press, 1977).

Matthews, Caitlín, *Sophia: Goddess of Wisdom* (London: Mandala, 1991).

Novalis, *Heinrich von Ofterdingen* (trsl. Palmer Hilty, *Henry of Ofterdingen*; New York: Frederick Ungar Publishing Company, 1972).

Novalis, *Hymns to the Night and other Selected Writings* (trsl. Charles E. Passage, The Library of Liberal Arts; Indianapolis, NY: Bobbs-Merrill Company, 1960).

———, *Pollen and Fragments* (trsl. Arther Versluis; Grand Rapids, MI: Phanes Press, 1989).

———, *Sacred Songs* (trsl. Eileen Hutchins; Spring Valley, NY: St. George Publications, 1978).

———, *Werke* ("Collected Works") (ed. Gerhard Schulz; Munich, Germany: C.H. Beck, 1969).

Paul, Lacquanna & Robert Powell, *Cosmic Dances of the Planets* (San Rafael, CA: Sophia Foundation Press, 2007).

———, *Cosmic Dances of the Zodiac* (San Rafael, CA: Sophia Foundation Press, 2007).

Ponente, Nello, *Who Was Raphael?* (trsl. James Emmons; Cleveland, OH: World Publishing Company, 1967).

Powell, Robert, *Christian Hermetic Astrology: The Star of the Magi and the Life of Christ* (Great Barrington, MA: SteinerBooks, 1998).

———, *The Christ Mystery: Reflections on the Second Coming* (Fair Oaks, CA: Rudolf Steiner College Press, 1999).

———, *Chronicle of the Living Christ: Foundations of Cosmic Christianity* (Great Barrington, MA: SteinerBooks, 1996).

———, *Hermetic Astrology, Vol. I: Astrology and Reincarnation* (San Rafael, CA: Sophia Foundation Press, 2007).

————, *Hermetic Astrology, Vol. II: Astrological Biography* (San Rafael, CA: Sophia Foundation Press, 2007).

————, *History of the Zodiac* (San Rafael, CA: Sophia Academic Press, 2007).

————, *The Most Holy Trinosophia: The New Revelation of the Divine Feminine* (Great Barrington, MA: SteinerBooks, 2000).

————, *The Mystery, Biography, and Destiny of Mary Magdalene: Sister of Lazarus John & Spiritual Sister of Jesus* (Great Barrington, MA: Lindisfarne Books, 2008).

————, *The Sign of the Son of Man in the Heavens: Sophia and the New Star Wisdom* (San Rafael, CA: Sophia Foundation Press, 2008).

————, *The Sophia Teachings: The Emergence of the Divine Feminine in Our Time* (Great Barrington, MA: Lindisfarne Books, 2007).

———— & Kevin Dann, *Christ & the Maya Calendar: 2012 & the Coming of the Antichrist* (Great Barrington, MA: Lindisfarne Books, 2009).

Ptolomy, *Tetrabiblos* (trsl. F. E. Robbins, Loeb Classical Library; Cambridge, MA: Harvard University Press, 1980).

Roder, Florian, *Novalis: Die Verwandlung des Menschen* ("Novalis: The Transformation of the Human Being") (Stuttgart, Germany: Urachhaus, 1992).

Ritter-Schaumburg, Heinz, *Novalis und seine erste Braut* ("Novalis and his first Bride") (Stuttgart, Germany: Urachhaus, 1986).

Scheps, Nick, *The Trutine of Hermes: A Guide to Calculating and Interpreting the True Ascendant* (Shaftesbury, England: Element Books, 1990).

Schipflinger, Thomas, *Sophia-Maria: A Holistic Vision of Creation* (trsl. James Morgante; York Beach, ME: Samuel Weiser, 1998).

Steiner, Rudolf, *According to Luke: The Gospel of Compassion and Love Revealed* (trsl. Catherine E. Creeger, CW 114; Great Barrington, MA: SteinerBooks, 2001).

————, *The Apocalypse of St. John* (trsl. Mabel Cotterell, CW 104; London: Anthroposophical Publishing Company, 1958).

————, *Christianity in Human Evolution* (CW 109; Spring Valley, NY: Anthroposophic Press, 1979).

————, *Esoteric Christianity and the Mission of Christian Rosenkreutz* (CW 130; London: Rudolf Steiner Press, 1984).

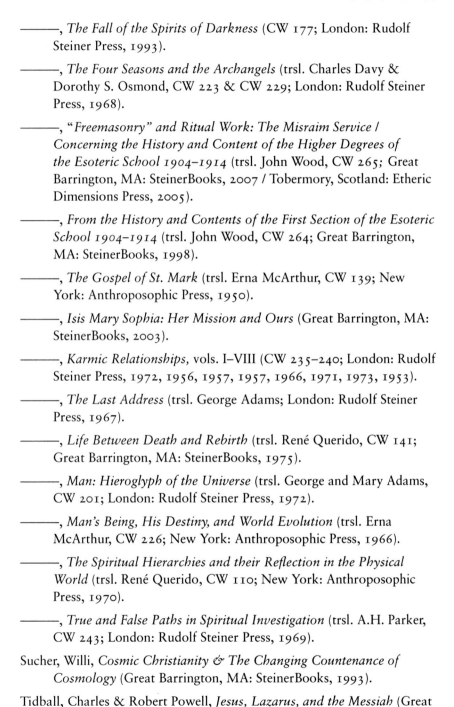

————, *The Fall of the Spirits of Darkness* (CW 177; London: Rudolf Steiner Press, 1993).

————, *The Four Seasons and the Archangels* (trsl. Charles Davy & Dorothy S. Osmond, CW 223 & CW 229; London: Rudolf Steiner Press, 1968).

————, *"Freemasonry" and Ritual Work: The Misraim Service / Concerning the History and Content of the Higher Degrees of the Esoteric School 1904–1914* (trsl. John Wood, CW 265; Great Barrington, MA: SteinerBooks, 2007 / Tobermory, Scotland: Etheric Dimensions Press, 2005).

————, *From the History and Contents of the First Section of the Esoteric School 1904–1914* (trsl. John Wood, CW 264; Great Barrington, MA: SteinerBooks, 1998).

————, *The Gospel of St. Mark* (trsl. Erna McArthur, CW 139; New York: Anthroposophic Press, 1950).

————, *Isis Mary Sophia: Her Mission and Ours* (Great Barrington, MA: SteinerBooks, 2003).

————, *Karmic Relationships,* vols. I–VIII (CW 235–240; London: Rudolf Steiner Press, 1972, 1956, 1957, 1957, 1966, 1971, 1973, 1953).

————, *The Last Address* (trsl. George Adams; London: Rudolf Steiner Press, 1967).

————, *Life Between Death and Rebirth* (trsl. René Querido, CW 141; Great Barrington, MA: SteinerBooks, 1975).

————, *Man: Hieroglyph of the Universe* (trsl. George and Mary Adams, CW 201; London: Rudolf Steiner Press, 1972).

————, *Man's Being, His Destiny, and World Evolution* (trsl. Erna McArthur, CW 226; New York: Anthroposophic Press, 1966).

————, *The Spiritual Hierarchies and their Reflection in the Physical World* (trsl. René Querido, CW 110; New York: Anthroposophic Press, 1970).

————, *True and False Paths in Spiritual Investigation* (trsl. A.H. Parker, CW 243; London: Rudolf Steiner Press, 1969).

Sucher, Willi, *Cosmic Christianity & The Changing Countenance of Cosmology* (Great Barrington, MA: SteinerBooks, 1993).

Tidball, Charles & Robert Powell, *Jesus, Lazarus, and the Messiah* (Great Barrington, MA: SteinerBooks, 2005).

Tomberg, Valentin, *Christ and Sophia: Anthroposophic Meditations on the Old Testament, New Testament, and Apocalypse* (trsl. R. H. Bruce; Great Barrington, MA: SteinerBooks, 2006).

———, *Inner Development* (trsl. Richard and Patricia Bloedon; Great Barrington, MA: SteinerBooks, 1992).

———, *Lazarus, Come Forth! Meditations of a Christian Esotericist on the Mysteries of the Raising of Lazarus, the Ten Commandments, the Three Kingdoms & the Breath of Life* (former title: *Covenant of the Heart*, trsl. Robert Powell and James Morgante; Great Barrington, MA: Lindisfarne Books, 2006).

Tresemer, David & Robert Schiappacasse, *Star Wisdom and Rudolf Steiner: A Life Seen through the Oracle of the Solar Cross* (Great Barrington, MA: SteinerBooks, 2007).

Vasari, Giorgio, *The Lives of the Artists* (Florence, 1550; trsl. Julia Conway Bondanella and Peter Bondanella, Oxford World Classic; Oxford, England: Oxford University Press, 1998).

van Emmichoven, J. E. Zeylmans, *A New Novalis?* (trsl. Elizabeth Spaey; Spring Valley, NY: St. George Publications, 1976).

von Dellingshausen, Erica, *Die Wartburg* (Stuttgart, Germany: Urachhaus, 1983).

von Keyserlingk Adalbert, (ed.), *The Birth of a New Agriculture: Koberwitz 1924* (trsl. John M. Wood; London: Temple Lodge Publishing, 1999).

———, *Monte Gargano: Europas ältestes Michaelsheiligtum* ("Monte Gargano: Europe's Earliest Sanctuary of St. Michael") (Stuttgart, Germany: Urachhaus, 1987).

Vreede, Elisabeth, *Astronomy and Spiritual Science* (trsl. Anne Riegel and Ronald Koetzsch; Great Barrington, MA: SteinerBooks, 2001).

Wachsmuth, Guenther, *Kosmische Aspekte von Geburt und Tod: Beiträge zur Karma-Forschung* ("Cosmic Aspects of Birth and Death: Contributions to Karma Research"; Dornach, Switzerland: Verlag am Goetheanum, 1990).

Computer Program:

Peter Treadgold, *Astrofire* (distributed by the Sophia Foundation of North America, Palo Alto, California). *Astrofire* has a comprehensive research module for data storage and comparison charts, a star catalog with over 4,000 stars, and a database of birth and death charts of historical personalities. It is capable of printing out geocentric and heliocentric/hermetic sidereal charts and ephemerides throughout history. With this program one can: compute birth charts in a large variety of systems (tropical, sidereal, geocentric, heliocentric, hermetic); calculate conception charts using the hermetic rule, in turn applying it for correction of the birth time; produce charts for the period between conception and birth; print out an "astrological biography" for the whole of life with the geocentric and heliocentric planetary systems; work with the sidereal zodiac according to the definition of your choice (Babylonian sidereal, Indian sidereal, unequal-division astronomical, etc.); work with planetary aspects with orbs of your choice. Included are eight house systems and a variety of chart formats. It also includes an ephemeris program with a search facility. *Astrofire* runs under Microsoft Windows. If you are interested in *Astrofire*, please contact the Sophia Foundation of North America, 3143 Avalon Court, Palo Alto, CA 94306, USA. Tel/Fax: (650) 494-9900. E-mail: sophia@sophiafoundation.org. Web site: www.sophiafoundation.org.

ABOUT THE AUTHOR

 ROBERT POWELL, Ph.D., is an internationally known lecturer, author, eurythmist and movement therapist. He is founder of the Choreocosmos School of Cosmic and Sacred Dance, and cofounder of the Sophia Foundation of North America. He received his doctorate for his thesis on the History of the Zodiac, now available as a book from Sophia Academic Press, and he is on the adjunct faculty of Wisdom University. His published works include: *The Sophia Teachings*, a six-tape series (Sounds True Recordings), as well as the following books: *Divine Sophia–Holy Wisdom*; *The Most Holy Trinosophia and the New Revelation of the Divine Feminine*; *The Sophia Teachings*; *The Mystery, Biography, and Destiny of Mary Magdalene*; *Chronicle of the Living Christ*; *Christian Hermetic Astrology*; *The Christ Mystery*; *The Sign of the Son of Man in the Heavens*; *The Morning Meditation in Eurythmy*; and the yearly *Christian Star Calendar*, as well as other works published by Sophia Foundation Press. He translated the spiritual classic *Meditations on the Tarot* and co-translated Valentin Tomberg's *Lazarus, Come Forth!*. Robert is co-author, with Kevin Dann, of *Christ & the Maya Calendar: 2012 & the Coming of the Antichrist* and, with Lacquanna Paul, *Cosmic Dances of the Zodiac* and *Cosmic Dances of the Planets*. He teaches a gentle form of healing movement: the sacred dance of eurythmy (from the Greek, meaning "harmonious movement"), as well as the cosmic dances of the planets and signs of the zodiac. Through the Sophia Grail Circle he facilitates sacred celebrations dedicated to the Divine Feminine. Robert offers workshops in Europe and North America and, with Karen Rivers, cofounder of the Sophia Foundation, leads pilgrimages to the world's sacred sites (1996, Turkey; 1997, Holy Land; 1998, France; 2000, Britain; 2002, Italy; 2004, Greece; 2006, Egypt; 2008, India). Web sites: www.sophiafoundation.org and www.astrogeographia.org.

CPSIA information can be obtained at www.ICGtesting.com
Printed in the USA
BVOW011834281112

306695BV00003B/751/P